A Translation of Luigi Paolucci's *On Birdsong*

This book is part of the Peter Lang Humanities list.
Every volume is peer reviewed and meets
the highest quality standards for content and production.

PETER LANG
New York • Bern • Berlin
Brussels • Vienna • Oxford • Warsaw

Luigi Paolucci

A Translation of Luigi Paolucci's *On Birdsong*

Phenomenology, Animal Psychology, and Biology

Translated and with an Introduction by Paolo Palmieri

PETER LANG
New York • Bern • Berlin
Brussels • Vienna • Oxford • Warsaw

Library of Congress Cataloging-in-Publication Data

Names: Palmieri, Paolo, translator, writer of added commentary.
Paolucci, Luigi. *Canto degli uccelli*. English.
Title: A translation of Luigi Paolucci's *On birdsong*: phenomenology,
animal psychology, and biology / translated and with an introduction by Paolo Palmieri.
Description: New York: Peter Lang, 2018.
Includes bibliographical references and index.
Identifiers: LCCN 2018007979 | ISBN 978-1-4331-4106-5 (hardback: alk. paper)
ISBN 978-1-4331-4099-0 (ebook pdf) | ISBN 978-1-4331-4104-1 (epub)
ISBN 978-1-4331-4105-8 (mobi)
Subjects: LCSH: Birdsongs. | Phenomenology. | Animal psychology. | Physiology.
Classification: LCC QL698.5.T73 2018 | DDC 598.159/4—dc23
LC record available at https://lccn.loc.gov/2018007979
DOI 10.3726/b10825

Bibliographic information published by **Die Deutsche Nationalbibliothek.**
Die Deutsche Nationalbibliothek lists this publication in the "Deutsche Nationalbibliografie"; detailed bibliographic data are available on the Internet at http://dnb.d-nb.de/.

29 Broadway, 18th floor, New York, NY 10006
www.peterlang.com

I dedicate this book to the
City of Ancona
and to the
splendor of the Marche

Table of Contents

Figures	ix
Preface	xi
Acknowledgments	xv
Note on Translation	xvii
Introduction: *The Human/Animal*	1
On Birdsong: An Essay on Animal Physiology and Biology in Relation to Sexual Selection and the Struggle for Existence by Luigi Paolucci	81
Part One: Comparative Phenomenology of Birdsong	83
Chapter One: On Natural Language	85
Chapter Two: The Mechanism of Sound in Autophonous Animals	101
Chapter Three: The Mechanism of Sound in Heterophonous Animals	115
Chapter Four: The Fundamental Sounds of Birdsong	125
Chapter Five: Musical Representation of Birdsong	135
Part Two: Patterns of Psychic Expression in Birdsong	145
Chapter One: Intelligence and Communication in Birdsong	147
Chapter Two: Diversity in Birdsong according to Age and Sex	157

Chapter Three: Biological Purposes of Birdsong 167
Chapter Four: Adult Birdsong in Both Sexes 181
Chapter Five: Imitation Voices 193
Chapter Six: Conclusion: Phonetic Corruption and Regeneration in Birdsong 197

Index 203

Figures

Figure I.1	*Portrait of a woman with a pearl* by Giovanni Agostino da Lodi.	16
Figure I.2	Man and his lion archetype.	19
Figure I.3	Spectrogram of the initial segment of the song.	41
Figure I.4	The spectrogram of a fragment of a nightingale song.	53
Figure 1.1	The frontispiece of Paolucci's monograph on birdsong.	84
Figure 1.2	A native man from Mallicolo.	93
Figure 2.1	Examples of cricket song notation.	105
Figure 2.2	Song of *Bufo Viridis* and shrill noise of *Hyla Viridis*	110
Figure 3.1	Bird vocal apparatus *Figure 1, Engraving III.*	120
Figure 5.1	Paolucci's music notation of the finch's song.	136
Figure 5.2	Examples 1–11 of birdsong as reported by Paolucci.	137
Figure 5.3	Examples 12–20 of birdsong as reported by Paolucci.	138
Figure 5.4	The raven picturing the stage of *nigredo* (blackness) within the alchemical vessel.	141

Preface

The naturalist Luigi Paolucci (1849–1935) was born in Ancona, the capital city of the Marche, a splendid region in central Italy facing the Adriatic. Not much is known of his life. He studied veterinary medicine in Ancona and then at the University of Bologna, where he graduated in 1870.[1] Paolucci spent most of his life in his city and the Marche, where he became a veterinary surgeon and a professor at local higher-education institutions. He was also marginally involved in political activities. He collected and prepared numerous animal taxidermies for a natural history museum, most of which are still preserved in Offagna, a lovely medieval town near Ancona.[2] Aside from his passion for natural history and birds, Paolucci had a great interest in languages, especially Greek, Spanish and Sanskrit. He dabbled in literature. He wrote poetry in the style of the Italian Romantics, perhaps for purely personal pleasure, as can be gathered from his own private diary. He especially admired Ugo Foscolo.

Paolucci's scientific and philosophical masterpiece is a monograph on birdsong, *Il canto degli uccelli*, published in 1878, which I have translated in this volume into English for the first time.[3] He also wrote on the Marche local flora and fauna and on sundry topics in natural history and other subjects. Perhaps it is no accident that Paolucci remains an elusive figure. He is virtually unknown in the records of the history of science. The paucity of biographical details and the scattered documents surviving in Ancona bear witness to this shadowy character.

The original manuscript of the essay here translated was recently discovered.[4] But there is a dimension of seclusion and intimacy in his life. Quotations of his work on birdsong are scant. His contemporary, Xenos Clark, an American naturalist, cited Paolucci's essay, in 1879, one year after the publication of "Il canto degli uccelli".[5] He wrote:

> This is an elaborate and philosophical memoir treating of animal song in many separate aspects. The notes of insects, batrachians, reptiles and birds are given in musical notation, and elaborately discussed. No less than twenty bird songs are thus written on the gamut, and the peculiar melodic quality of sixty-eight more tabulated. We hope to review this important work in a succeeding number of the Naturalist.[6]

The brief paragraph contains acute remarks. Clark intuited the scientific and philosophical ambition of Paolucci's essay. Unfortunately, the American naturalist did not follow up on his comments, and so far as I have been able to ascertain the promised review was never written or published. Aside from Clark, I could not find a single citation of Paolucci's essay in the modern scientific literature on birdsong and/or animal music. Paolucci was forgotten. Yet his essay was ground-breaking and in many respects has remained unrivalled after more than a century. Most modern research on birdsong has been done in typical academic settings, which condition the mindset of the scientists and limit their approach within narrow materialistic-mechanistic prejudices. In sharp contrast, Paolucci's work matured in a life-world in which his experience of dwelling in the ecologies of birdsong, immersed in the pristine landscapes of the Marche and the Adriatic, was a form of humble involvement with nature, in reverence of the mystery of animal life. It was this tension of renunciation, fascination with the animal 'other' and mystical exaltation, which transpires so genuinely from the pages of his essay, that inspired me when I encountered his magical work on birdsong.

In my introductory essay, *The human/animal*, I propose a meditation on the question that captured my imagination when I first read "Il canto degli uccelli". It is the question of the demarcation of the realm of the human from that of the non-human animal. I was struck by the fact that Paolucci extensively and systematically employs musical notation to transcribe patterns of birdsong. This was not totally new since others before him had occasionally adopted this practice. But I found it intriguing that the musical notation of birdsong had opened up for Paolucci (and for myself) a new vista on the presence of music in the non-human animal kingdom. Thus, my introductory essay weaves many philosophical themes that are historically informed, and which gravitate towards the question of the demarcation of the human from the non-human animal. Examples include non-human animal reason, the ethical issues concerning non-human animals, the uncanny parallelisms between human and non-human languages and laughter, the acoustic interpretation of birdsong patterns, psychoacoustics, and the neurophysiology of hearing in humans and birds. There is no strong unifying motive, however. I am a nomadic thinker, an anarchist scavenger who loves feeding on intellectual carcasses in the margins of culture.

In the notes to Paolucci's essay, I assembled scholarly, sober comments on the literary sources and the technical problems that Paolucci must have faced when researching and writing the essay. He relied on a vast naturalistic and philosophical literature. Occasionally, my annotations become more personal, and I launch into extemporary ramblings. However, much remains to be done to honor the lesson that Paolucci has to teach still today, at the beginning of the twenty-first century. I hope that this translation will motivate others to read the sources and penetrate deeper into the questions that I raise in the introductory essay, in conversation with a motley crew of philosophers, poets, and free-thinkers.

Notes

1. On the intellectual climate at the University of Bologna, during a period spanning the second half of the nineteenth century to the early twentieth century, cf. the many essays edited in Odrini and Tega 1990, which offer valuable discussion on the relation between science and philosophy and especially the philosophical reception of Darwinism at Bologna and in Italy.
2. Mangani 1996, pp. 8 ff, and 1982, pp. 83–90. Cf. also the excellent interpretive essays by various authors collected in Mangani 1982.
3. Paolucci 1878.
4. A valuable collection of documents including the manuscript, the private diary and letters to and from Charles Darwin is preserved in the *Bilbioteca Francescana*, in Falconara Marittima, Ancona. Giorgio Mangani was able to track down the collection in 2012. An analysis of the precious manuscript remains for future research.
5. Clark 1879.
6. Clark 1879, p. 215. It is remarkable that Paolucci's essay was known to the American naturalist, and that an American scientist should still read Italian in the late nineteenth century. Unfortunately, the imperialistic ascendance of English in the twentieth century as the only valuable scientific language has dramatically impoverished the scientific community and turned science into a subtle but vicious form of neocolonialism.

Bibliography

Clark, Xenos. 1879. "Animal music, its nature and origin". *The American Naturalist* 13(4), pp. 209–223.

Mangani, Giorgio. 1982. [Ed.] *Mostri e fossili. Il gabinetto di storia naturale di Luigi Paolucci.* Ancona: il lavoro editoriale.

———. 1996. "Il naturalista Luigi Paolucci (1849–1935)". In: Piazzini 1996, pp. 7–13.

Odrini, Guido, and Tega, Walter. 1990. [Eds.] *Filosofia e scienza a Bologna tra il 1860 e il 1920.* Bologna: Cappelli Editore.

Paolucci, Luigi. 1878. *Il canto degli uccelli. Note di fisiologia e biologia zoologica in rapporto alla scelta sessuale e alla lotta per l'esistenza.* Milan: G. Bernardoni. Retrievable online at: http://dfgviewer.de/show/?set[mets]=http%3A%2F%2Fwww.zvdd.de%2Fdms%2Fmetsresolver%2F%3FPPN%3DPPN641658559&set[image]=1

———. 1882. "Il canto degli uccelli. Note di fisiologia e biologia zoologica in rapporto alla scelta sessuale e alla lotta per l'esistenza." *Rivista di Filosofia Scientifica,* Anno Secondo, Volume 2, pp. 297–320, and pp. 650–682.

Piazzini, Stefano. 1996. [Ed.] *Le collezioni Paolucci di storia naturale.* Ancona: il lavoro editoriale.

Acknowledgments

It is a great pleasure to acknowledge the extraordinary hospitality that I received when I visited Ancona for research purposes in 2012. Giorgio Mangani, the foremost Paolucci scholar and a fine student of the history and philosophy of cartography, helped me immensely in gathering the sparse information on Paolucci that was available in libraries and museums. On the occasion of my visit he discovered the original manuscript of the essay on birdsong and kindly shared his knowledge with me. Thanks to Giorgio Marinelli who patiently photographed the manuscript in high resolution. Father Giancarlo Mandolini, Library Director, generously donated his time to allow me to study rare materials in the *Biblioteca Francescana* in Falconara Marittima (on the outskirts of Ancona), where the surviving documents of the Paolucci family are preserved. I thank him warmly from the heart. Many thanks to Alfonso Capriolo who welcomed me friendly, during my stay in Ancona, in his capacity as Director of the Museum System of the Marche. I have fond memories of a spectacular lunch *en plein air* at Offagna (where Paolucci's museum of animal taxidermies is permanently exhibited), with friends and local city officials. Last but not least I wish to remember a fabulous dinner with Giorgio Mangani and Cristina Morbiducci-Mangani, two exquisite persons who shared with me culinary insights, Marche folklore, political hopes and academic despair, and who encouraged this project from the very beginning. Above all, this work was inspired by the loving care of my wife Paola without whose support and presence it would not have been possible. Finally, I am very grateful to the many students who followed my courses over the last few years at the University of Pittsburgh, and especially to the daring youths who participated in a historic, multi-disciplinary graduate seminar, entitled *the human/animal*, which laid the philosophical and pedagogical foundations for the introductory essay in this volume.

Note on Translation

I had no idea what translation really means before undertaking this project. After struggling with Paolucci's essay for a long time, I realized that translation is a transformative process in which both the translator and the author enter into a strange pact of allegiance. Both will suffer at each other's hands. I hope I made justice to the clarity and elegance of Paolucci's language. He was a superb stylist of scientific prose. His command of the Italian language was impeccable and precise, the syntax of the essay is fluid and the lexical variety admirable. I have done my best to translate the ornithological nomenclature accurately though discrepancies may occasionally be expected when Paolucci employs common, or idiomatic names which might not overlap satisfactorily with the current classification of species. I distinguished my annotations to the essay from Paolucci's own notes by placing mine between braces. I based my translation on the text published in 1878, a copy of which is in my personal library.[1]

Note

1. Paolucci 1878. Republished in Paolucci 1882. See Bibliography to Preface.

Introduction: *The Human/Animal*

Le mal de la souffrance—passivité extrême, impuissance, abandon et solitude—n'est-il pas aussi l'inassumable et, ainsi, de par sa non intégration dans l'unité d'un ordre et d'un sens, la possibilité d'une couverture et, plus précisément, de celle où passe une plainte, un cri, un gémissement ou un soupir, appel originel à l'aide, au secours curatif, au secours de l'autre moi dont l'altérite, dont l'extériorité promettent le salut ?[1]
Emmanuel Levinas, *La souffrance inutile*, quoted from: *Entre nous. Essais sur le penser-à-l'autre* (1991).

1. The Human/Animal *Visage*

Levinas's phenomenology of suffering accentuates the *uselessness* of pure suffering, the suffering that cannot be *assumed* in, or *subsumed* under, a structure of meaning, namely, an order in which the experience of the sufferer would appear in the light of a horizon of purposes through which suffering would be justified, understood, and ultimately accepted. Levinas illustrates the theme of useless suffering by arguing that its archetypical, or paradigmatic manifestation, as he put it, must have been the Holocaust of both the Jewish population and other ethnic and social groups in the Nazi concentration camps. The disproportion between useless suffering and divine justice (which Levinas calls 'theodicy') manifested itself at Auschwitz in a self-evident manner, where God remained conspicuous for its inexplicable absence. Levinas's phenomenological essay on useless suffering, which originally appeared in 1982,[2] was fleshed out later on by Primo Levi, in a narrative of dramatic depth, which gave Levinas's analysis—a rather restrained if not sanitized account—the horrific vividness of the daily reality of the sufferers in the flesh as they had been uselessly dehumanized and massacred at Auschwitz. Himself an Auschwitz survivor, in his last book, "The submerged and the saved", published in 1986, the year before committing suicide, Levi wrote a fundamental chapter entitled 'Useless violence', whose title may have been inspired by Levinas's essay (though Levi never quoted the French philosopher).[3]

Useless dehumanization begins, in Levi's narrative, with the transformation of an ordinary means of transportation, the train, into an instrument of torture and death. The train that carries the prisoners to Auschwitz shows the monstrous that is latent in the ordinary. Once the seats have been removed and the wooden floors laid bare, the train cars carry up to about a hundred people (but no more than fifty or so in the case of those leaving from Italy), who can only stand, which makes the trip a horror (an 'inferno') as there is no sanitation, toilettes, or any possibility of lying. Nothing to drink or eat is given to the people, whom are not even so much as alerted by Nazi officers that they should bring their own food and water. This deliberate omission, Levi notes, is the first example of useless violence. In his train, Levi recounts, the lack of a toilette caused a much greater suffering than cold and thirst. The trauma of defecation in public was unbearable for many people, and it was a trauma for which European civilization seemed utterly unprepared. It was, for Levi, the ominous signal of a gratuitous yet deliberate, malignant scheme. During a stop in a transit station in Austria, the SS guards that accompanied the prisoners laughed at seeing so many people rest as best as they could on the curbs of sidewalks, or on the railroad. German travelers openly expressed their disgust for those *beings* that infested the station with strange patterns of behavior. "They are not human beings, they are not *Menschen*, but beasts, pigs".[4] They deserve their fate. In the camps, prisoners needed to habituate quickly to the ritual of defecation in public. Levi comments that such habituation was a sure sign of the effective transformation of the prisoners from human beings into non-human animal beings. But, says Levi, it was never a conscious decision on the part of the bureaucracy that managed the camps system. The useless cruelty of the violation of personal care and dignity was systemic in the camps. Yet, it was the result of the intrinsic logic of the camps system. An organizational system that dehumanizes the human must be efficacious in all directions including those that affect all aspects of daily life and personal care. Analogous to the violence of defecation was that of nudity. Daily life was constellated by innumerable occasions that obliged the prisoners to expose their naked bodies to guards and/or medical officers. There might have been a minimum of logic in all that, says Levi, hinting, for instance, at the medical examination for lice, yet the redundancy of the practices was surely useless and offensive. The fact is, Levi insists, that the human being that is forced to take their clothes off, and hence is exposed in their defenseless nudity, perceives themselves no longer as humans but as a worms, low, ignoble, which can be crushed at any time. In most of its obsessive patterns of order-for-the-sake-of-order, the camps system was, according to Levi, the translation of German, long-established military practices of discipline and orderly lifestyle in the barracks. The army of prisoners in the camps must be a caricature of the German army. The ritualized drill of making the bed in the dorms and the consequent punishment for failing to respect the standards of alignment of the blankets and cushion were emblematic of this caricature. They carried the logic of the system to the extreme of violence-for-the-sake-of-violence. The forced tattooing of prisoners gave the experience of suffering its seal of irreversibility. It was the branding of slaves and cattle. Mosaic law prohibits tattooing for orthodox Jews, a sign of barbarism, and thus the practice became even more dehumanizing for them. Making the prisoners suffer was a daily obligation for the Nazi officers in the camps. It was not a question of reasoning about it, there were no other motivations. But, for Levi, the perpetrators were not made of a different *substance* than that of other human beings. They were not different than us. They had been conditioned to think and act in that manner by a system of coercion, education, and propaganda that in the totalitarian state put ordinary morality upside-down. In this crescendo of the torture of the flesh, I wish to recall Levi's pointed observation that a disturbing, glaring example of purposeless suffering was the profanation of the human body turned into an object of meaningless experimentation, an object of manipulation, a thing that no longer belonged to anybody and which was at anybody's disposal.

At this point Levi's narrative is struck by the blows of an obscure noise accompanying a funeral in the mind. A symptom of silence is thrown up against us, the readers, that provokes a rupture in the tissue of our thoughts. Levi emphasizes that the horrific transformation of the human body into a manipulable thing in the Nazi concentration camps must be evoked in the context of the emerging awareness, in society at large, that the use of non-human animals in scientific experimentation can and should be questioned.[5] It is the question of useless suffering, explored by Levi with an admirable balance of passionate participation and rigorous analytical exposition, that provokes the casting of a shadow into his and our consciousness. It is the shadow of the presence in our laboratories of the non-human animal, the sufferer that has agreed to keep silent about its suffering for purposes of research in the light of human ideals. Thus, the pursuit of scientific knowledge, violence, ethics, and useless suffering appear to expose a repressed nexus of intricate relationships, philosophical denial, and social responsibilities in the history of modern science and throughout Western civilization.

In the remainder of this section and beyond, I will occasionally be inspired by Levinas's philosophy of the face (French *visage*) as foundation for an ethical way of life. I will search for principles that might create a crisis in the epistemic certainty that legalizes useless non-human suffering for human purposes. However, I will depart from his thoughts by radically rethinking the *face* (for Levinas, always an individual *human* face, namely, *le visage d'Autrui*). I will proceed by imagining myself at the *human/animal* frontier, that is, at the threshold (indicated by the slash character in the compound neologism *human/animal*) where conceptual demarcation between species breaks down, taxonomy becomes pretend-play, and where topographic boundaries cannot be drawn except by conquer and violence. A borderland where human speech returns to the cradle of non-human animal vocalization and rhythm and song. I will imagine the human/animal face in the seamless realm of being where birdsong disrupts the silence of beings. I will make use of philosophical language, ornithological metaphors, and worn-out categories to break through the flatland of Western philosophical discourse. I will haul the detritus away, in order to pierce into a three-dimensional imaginative space and lose myself and then mount an impossible rescue operation riddled with sin, compromise, and logical paradox.

My starting point is 'biosophical' in the sense of reminding ourselves that the living imaginative organism is implicated in the processes of the imagination, and that the separation of the living organism from the imagination and its productions must be suspended, bracketed. The biosophical act of imagining must aim at keeping the whole organism within its horizon, together with her history, travails, reveries and labors. Thus I begin by recollecting a childhood memory of useless violence.

A living pigeon was given as a present to my mother by a relative who had a small butchery business. My mother had the bird gently covered in a blanket and then placed it in the tub in our bathroom. I had seen the bird in the tub, alive and well, in deep silence, plunged into an infinite solitude, clutching calmly at its being alive, first with curiosity and then immediately with irresistible sympathy, a deep sense of empathy was filling me, as I stared at its face, as it stared at me, inviting help from the other "whose alterity, whose exteriority promises salvation". Later in the evening, when my mother entered the bathroom to kill the bird, I took refuge in my bedroom and cried my heart out and eloquently. I remember my mother being taken aback by such an explosive reaction. Perhaps she tried to concoct something of an explanation, or consolatory gesture to my wounded self before she proceeded—uneasily, or so it seemed to me many years ago—to kill the bird anyway, fatefully, with righteousness.

Le Visage d'Autrui

There are tantalizing etymological issues concerning the word '*visage*'. It is not a matter of linguistic archeology and philological exactness. Rather the ancient roots of words tend to expose deeper strata in the animal consciousness. They are concomitants of the early stages of primate and hominid evolution like tools invented for the work of the imagination. *Visage* comes from the Latin *vultus*, which in turn does not seem to have a cognate in Greek philosophical parlance but might instead be connected with Sanskrit origins, specifically the verbal root meaning *choose, desire, will.*[6] A controversial claim in this regard was already made in antiquity by Cicero to the effect that the Greek language does not have an equivalent term for *vultus*. He asserted that

> with respect to man, this same bountiful nature has not merely allotted him a subtle and active spirit, but also physical senses, like so many servants and messengers. And she has laid bare before him the obscure but necessary explanation of many things, which are, as it were, the foundation of knowledge; and in all respects she has given him a convenient figure of body, suited to the bent of the human character. For while she has kept down the countenances of other animals, and fixed their eyes in their food, she has bestowed on man alone an erect stature, and prompted him to the contemplation of heaven, the ancient home of his kindred immortals. So exquisitely, too, has she fashioned the features of the human face, as to make them indicate the most recondite thoughts and sentiments. For our eloquent eyes speak forth every impulse and passion of our souls; and that which we call 'expression' [*vultus*], which cannot exist in any other animal but man, betrays all our feelings, the power of which was well known to the Greeks, though they have no name for it.[7]

The natural candidate for a Greek word equivalent to *vultus* would be *prosōpon* (from which, for example, we derived *prosopopoeia*). Yet Cicero claims that Greek does not have a noun for the Latin *vultus*. Already the French humanist Adrien Turnèbe, in the sixteenth century, commented that although a Greek word was indeed available (*prosōpon*), what was missing in the Greek language was the denotative force of the word *prosōpon*, which, according to him, did not have the semantic element of will inherent in the Latin *vultus*.[8] Turnèbe seems to have intuited what later philologists (such as Ottorino Pianigiani[9]) would suggest about the earlier Indo-European origin of the Latin *vultus*. It is this strong element of will (desire, the striving for), inherent in *vultus*, that should not be too hastily denied to the non-human animal. Instead it should trouble us, for it raises the dramatic question whether we should recognize a *visage* in the non-human animal face that comes forward towards us, the face that wills, desires, or supplicates, as, for example, the face of the pigeon that I remember vividly, in my imagination, when it was on the brink of the abyss from where it invoked the exteriority that promises salvation. Cicero separated the countenance of the non-human animal, deprived of expression and fixated in an eternal gesture downwards forever intent on ingesting food, from the face of the human aiming upwards at the heavens where the Gods have their abode. Levinas has put forward the theme of the human *visage* with great force, making it the center of gravity of his philosophy, or the focus of his imagination. But the *visage d'Autrui* remains captive of the Ciceronian prejudice promulgated throughout Western thought ever since antiquity, which drastically demarcates the human from the non-human animal. "The *visage* does not resemble the plastic form, always already deserted, betrayed by the being that it reveals, like the marble from which the gods that it makes visible have already departed. [The epiphany of the *visage*] differs from the animal face where being, in its brute stupidity, does not encounter itself".[10] I concur with Jacques Derrida's view that

> even if Levinas inflects what he inherits, even if he inverts what could be described as the traditional and ontological tendency concerning the subject, even if he does that in a strong, original, and, let's say, subversive manner [...], even if he submits the subject to a radical heteronomy [...], even if says about the subject that it is above all a 'host' [...], even if he reminds us that the subject is a 'hostage' [...], this subject of ethics, the face, remains first of all a fraternal and human face.[11]

We must imagine otherwise than human, and instead pursue the paradox of the human/animal *visage*. I will follow a Levinas thread in and out of the labyrinth, despite Derrida's critique, for I think that I can find in Levinas's profound and original thinking about the human *visage* the *plusquam*-philosophical resources that will guide me beyond the encircling walls of dialectical humanistic reasoning, which even Derrida's meanderings in pursuit of the animal do not seem to escape.[12]

Before I turn to that, however, I will mobilize figures of the imagination from subtly interrelated scenarios. First, I will return to the mysterious Greek *prosōpon*, and investigate the question whether, despite its falling short of forcefully denoting the will, and despite the inadequacy implied by Cicero's denial, *prosōpon* might still suggest analogies for imagining otherwise than human, for exploring the beguiling paradox of the non-human animal face, and hence for refashioning the *visage d'Autrui* into that impossible singularity of the human/animal that comes before the violent demarcation of the domains of being. Secondly, I will hybridize the Geek background with the Judeo-Christian imagination of will and *vultus* in Dante's *Divine Commedy*, more precisely, in the first Cantos of his *Paradise* where a surprisingly effective poetic anamnesis is given of the relation between freedom, will, and *vultus*. Thirdly, I will follow the adventure of the physiognomic imagination at the threshold of European modernity. I will take a stroll into the natural-magical realm of the face and the body as expressions of a human/animal typology of character. A cognitive adventure that I can undertake safely, as Levinas would prescribe,[13] without subsuming the exteriority of the absolutely other into the annihilating imperialism of my subjectivity, but at the same time without my succumbing to the heteronomy of an infinite exteriority, exercising a dominion that would suffocate my freedom by devouring my own self in a mystical embrace.

The Face on the Disk of the Moon[14]

Perhaps the most intriguing context for the discussion of *prosōpon* and its relation to the enigma of the face is Plutarch's dialogue *On the face of the moon*. It influenced modernity throughout, shaping its scientific and philosophical reforms, modeling some of its most enduring intellectual structures, and fuelling the cosmological revolution that replaced the geocentric model of the cosmos with the Copernican system of the infinite universe (Johannes Kepler translated into Latin and commented the entire dialogue, and Galileo knew it well). The emergence of the autonomy of modern subjectivity cannot be dissociated from the solar thinking that placed the sun at the center of the universe. It is indeed its most glaring manifestation. Heliocentrism, especially the mechanistic framework that ascended to prominence with Descartes, culminated during the Enlightenment and which lasted well into the early nineteenth century, became the major force of modern philosophical thought. It violently disrupted non humanistic forms of the imagination such as Renaissance magic, physiognomy, laboratory and spiritual alchemy, and Hermeticism. It is striking that its corroding presence in modern philosophy and in many postmodernisms must have persisted for so long, throughout the entire twentieth century, and well after the decline of the classical mode of scientific thinking that the quantum revolution of the late 1920s and 1930s brought about. Phenomenology and existentialism did not escape this bizarre conundrum. The fossil survival of Copernican subjectivism and mechanistic modes of philosophizing in both Husserl's phenomenology of objects and Heidegger's phenomenology of the world (deeply rooted in his staunch scholastic Catholicism) is all too evident to anybody who is acquainted with the outlines of the history of Western science and its devastating influence on modern philosophy. A biosophical approach to the face recovers ancient and long-forgotten pathways to cosmic wisdom in order to overcome the mechanistic framework and the Copernican subjectivism that have pervaded modern philosophy.

To all intents and purposes, Plutarch's dialogue has much more to offer to a biosophy of the human/animal face than has hitherto been suspected. Hence my discussion is not to be construed as a philologically ambitious interpretation but as a creative reconstruction aimed at evoking archetypes emerging when the enigma of the face comes to the fore in historically-informed philosophical discussion. The word *prosōpon* also refers to the mask that ancient actors wore in the theater. It is therefore connected with their speeches resonating through the large aperture of the mouth. It is also connected with the wide-open apertures of the eyes that allowed the actors to see through the mask. But of course the mask does not have a range of facial expressions. It comes with a fixed countenance which seems to have been typified according to epoch, genre, styles, and places.[15] This fixed countenance is like the face on the disk of the moon as it appears to the motley characters of the dialogue. Their colorful speeches about the moon's face take us back to an imaginative epoch of mind when science and philosophy and myth had not yet been arbitrarily separated. The importance of the ethnic diversity of the cast of characters for the understanding of their speeches should not be underestimated. The leader of the dialogue is Plutarch's brother, Lamprias, and he is also the narrator. His Greek name evokes the quality of clarity and euphony. He was an archon of the oracle at Delphi, and is described as speaking loudly, well capable of concocting a story to lend credence to an argument, as a connoisseur of fine foods and dance, and also as somebody who will not play the joy-killer among young men. He narrates an earlier discussion in which he took part with friends, which in turn resumed and summarized an earlier conversation where Plutarch himself had been present in a leading role. The two most ambiguous characters (together with Lamprias) are a Carthaginian, Sulla, and a Roman-Etruscan, Lucius. Thus, we are confronted with three distinct ethnicities, the Greek, the Carthaginian, and the Roman-Etruscan. These characters carry into the narrative three inflections of the imagination. The Greek one is connected with logos, which in some sense is rationalism (often syllogistic argumentation), the Carthaginian is connected with mythology, as we shall see, for Sulla is responsible for contributing the eschatological myth that brings the dialogue to a closure almost leaving the reader in state of suspended animation, and the Roman-Etruscan is connected with the sect of the Pythagoreans (as Lucius is supposed to have been the pupil of a representative from that school, even though it remains rather mysterious what membership in this philosophical current or religious sect might have implied in the cultural cauldron of the Roman empire of the first century AD). Moreover, these ethnic diversities continually melt down into the speeches so as to give rise to dynamic formations that escape easy categorization and dissolve boundaries. It is clear that the narrative is constantly subjected to the strongest imaginative tension created by the principal and sole narrator of the dialogue who, let us not forget, is capable of excogitating stories to prop up the arguments. At all times we must guard against taking what is being said at a conceptual face value.

The moon displays a face in chiaroscuro which, as Lamprias demonstrates, is not the mirror-like reflection of earthly features such as the continents and the great ocean. The moon is not a mirror of the earth. The moon's apparent face, in other words, cannot be entirely appropriated by the human imagination. It cannot be conceptualized by thinking from the terrestrial viewpoint of discursive subjectivity. The face is beyond the reach of reflective thinking.[16] Chiaroscuro is deceptive. It creates the illusion of three-dimensionality on a flat surface. Its technique was perfected by Renaissance artists such as Leonardo, a sublime conjurer of both the human and the non-human face (recall the glorious *Woman with an ermine*). But it presupposes a flat surface operating like a mirror, on which light falls from a particular direction (or more than one) and bounces off towards the eyes of the viewer. It also presupposes the attention of the viewer who compasses the play of shadows in order to reconstruct the three-dimensional features of the image in the space of the imagination. This *viewer* of the face must absorb the light reflected from the surface in order to imagine the volume. He is guided by autonomous principles of order whose workings are unknown. *Imagining* as opposed to *viewing* the face requires imagining beyond the illusions of chiaroscuro on flat surfaces. It need not absorb light in an autonomous process of assimilation. It just requires learning to observe light falling on terrestrial objects, three-dimensional objects, capable of opposing resistance, opaque surfaces that cast shadows by blotting out the rays of light from the sun impacting the surfaces. This analogy of light falling on three-dimensional objects shapes the organic imagination of the face. Facial features turn into cavities and crevices and protuberances giving relief and volume to a landscape. Not an image, not a reflection, rather the face on the moon opens up a vista on a mysterious, cosmic landscape whose cosmological features appeal to the living organism for imaginative exploration of its environment.

To the 'absurd' Stoic doctrine according to which the moon is a mass of air mixed with fire, Plutarch opposes the theory of the moon like another earth (more precisely, a mixture of earth-like matter and star) or rather an extension of the terrestrial realm into the region of the moon.

> It is in fact not incredible or wonderful that the moon, if she has nothing corrupted or slimy in her but garners pure light from heaven and is filled with width, which is fire not glowing or raging but moist and harmless and in its natural state, has got open

> regions of marvelous beauty and mountains flaming bright and has zones of royal purple with gold and silver not scattered in her depths but bursting forth in abundance on the plains or openly visible on the smooth heights. If through the shadow there comes to us a glimpse of these, different at different times because of some variation and difference of the atmosphere, the honorable repute of the moon is surely not impaired nor is her divinity because she is held by men to be a celestial and holy earth rather than, as the Stoics say, a fire turbid and dreggish.[17]

It is a celebration of the intermediate position of the moon, in proportion to autonomy and heteronomy, as an organ of the cosmic animal that does not privilege a solar subjectivism of the human.

> Fire, to be sure, is given barbaric honors among the Medes and Assyrians, who from fear by way of propitiation worship the maleficent rather than the reverend; but to every Greek, of course, the name of earth is dear and honorable, and it is our ancestral tradition to revere her like any other god. As men we are far from thinking that the moon, because she is a celestial earth, is a body without soul and mind and without share in the first-fruits that it beseems us to offer to the gods according to custom requiting them for the goods we have received and naturally revering what is better and more honorable in virtue and power. Consequently let us not think it an offence to suppose that *she is earth and that for this which appears to be her face, just as our earth has certain great gulfs, so that earth yawns with great depths and clefts which contain water or murky air*; the interior of these the light of the sun does not plumb or even touch, but it fails and the reflection which it sends back here is discontinuous.[18]

The allegedly pure *Greek* character of the imagination of the earth and of the religion of the divinity of the earth-like moon, the mystery of her *prosōpon*, as opposed to the solar superstition of Medes and Assyrians—though riddled with racist prejudice and paradox—points to the definition of the *biosophical* imagination as an earthy, feminine space of meditation and equilibrium, harmony, even repose, between the hypertrophy and autonomy of solar subjectivity and the engulfing heteronomy of cosmic Stoic reason, here parodied as barbaric solar mysticism.

The mystery of the face/*visage* of the moon calls for recognition of her *unnatural* constitution. Surprisingly, it is this very unnaturalness of *visage* that flings open the gates of a social and ethical space for community and justice to enter the scene and settle in. The Stoic position that Lamprias's imagination rejects vehemently presupposes the notions of natural place and of the concomitant motion of all bodies in the cosmos towards their natural places. This peculiar idea of order rules out the possibility of the earthiness of the moon as being unintelligible, that is, beyond the reach of logos because *dis-placed*, because it is out of her natural place, no longer abiding cosmic logos, as the only natural place for earthy things to be must be the earth, which is sitting at the center of the cosmos and offering itself there as the abode of all things heavy by occupying an imaginary, immaterial point that is called 'center'.[19] If

> Earth had no part in heat, water no part in air; if there was not anything heavy above or anything light below; but the principles of all things were untempered and unamiable and solitary, not accepting combination or association with one another, but avoiding and shunning one another and moving with their own peculiar and arbitrary motions, then things would be in the state in which, according to Plato, everything is from which God is absent, that is to say in which bodies are when mind or soul is wanting. So they were until desire came over nature providentially, for Affection arose or Aphrodite or Eros, [...] in order that by changing position and interchanging functions and by being constrained some to motion and some to rest and compelled to give way and shift from the natural to the better, [beings] might produce a universal concord and community.[20]

When desire came over nature, a new order was set upon things that established community and concord, that is, justice. The unnaturalness of the constitution of things that have been squandered about by the supervenience of desire is what measures desire. The ratio of 'disorder' in the cosmos hints at the savvy labors of desire that founded community and justice. If

> we are finally to throw off the habits and opinions that have held our minds in thrall and fearlessly to say what really appears to be the case, no part of a whole all by itself seems to have any order, position, or motion of its own which could be called unconditionally 'natural'. On the contrary, each and every such part, whenever its motion is usefully and properly accommodated to that for the sake of which the part has come to be and which is the purpose of its growth or production, and whenever it acts or is affected or disposed so that it contributes to the preservation or beauty or function of that thing, then, I believe, it has its 'natural' position and motion and disposition.[21]

Hence nature is biosophical in the sense of preserving the thing or its functionality, and at the same time it is esthetic in that it contributes to the beauty of the thing. But nature does not contribute to justice or community. The latter function is a higher principle than nature. It is the biosophical principle of desire, the workings of Affection or Aphrodite or Eros. In this cosmos or community of things regulated by unnatural desire and higher concord, there is no role for speciesism and/or any racisms. These are phantoms of the categorical mind that came into the world with modern heliocentric subjectivity, and which reduced nature to the mechanistic universe and life to the materialistic formations of atoms and molecules. In the community of unnatural desire things have absolute value and uniqueness. Earthy things find hospitality and repose on the earth, not because earth is their *natural* place to come to rest, as the Stoics would have it, but because community establishes itself everywhere in the cosmos according to justice, and hence earthy places in the cosmos turn out to be the just place to live in for earthy things. Every being in the cosmos is the unique incarnation of the same justice in a different destiny. Every *visage* in the cosmos is the unique manifestation of the same infinity in an absolute, singular otherness.[22]

Levinas insisted that the relationship of the thinking subject to the idea of the infinite is not that of contemplation. The infinite is not an object of contemplation for a contemplative subject.[23] The question remains open of the mode of grasping the infinite as a possible object for the subjectivity. But perhaps the question is already entangled with the assumption that subjectivity plays a non-contemplative yet grasping role towards the idea of the infinite that has been implanted in it, and from which it has received an inflation, an excess that does not stem from within itself. According to the notion of autonomous subject that has prevailed in modern philosophy (in Levinas's view) the subject itself would be the author of its own inflation. In his account the inflation in the subject is brought about by the idea of the infinite implanting itself into the subject. However, both Levinas's and the modern approach (Cartesian contemplation[24]) seem to fail to reach the delicate balance preventing that either total autonomy preys on the infinite object or the total heteronomy of the infinite engulfs the finite subject. What is crucial, to my mind, is that although *visage* brings the infinite into grasping contact with the thinking subject, it cannot allow its infinity to become an object of the subject's contemplation. Rather *visage* must operate like a mediating entity in the relationship between finite and infinite. It is the moon in her eschatological role that casts light on this delicate mediating role. The eschatological myth recounted in the last part of Plutarch's dialogue brings this extraordinary function of the imagination of the face of the moon into clear focus. We can imagine otherwise than human by dreaming like the souls on the moon.

Here is the myth of the double death of humans.

> Most people rightly hold man to be composite but wrongly hold him to be composed of only two parts. The reason is that they suppose mind to be somehow part of soul, thus erring no less than those who believe soul to be part of body, for in the same degree as soul is superior to body so is mind better and more divine than soul. The result of soul and body commingled is the irrational or the affective factor, whereas of mind and soul the conjunction produces reason; and of these the former is source of pleasure and pain, the latter of virtue and vice. In the composition of these three factors earth furnishes the body, the moon the soul, and the sun furnishes mind to man for the purpose of his generation even as it furnishes light to the moon herself. As to the death we die, one death reduces man from three factors to two and another reduces him from two to one; the former takes place in the earth [...], the latter in the moon....[25]

But what is the fate of soul after the second death? "The substance of the soul is left upon the moon and retains certain vestiges and dreams of life. [...] The soul receives the impression of its shape through being molded by the mind and molding in turn and enfolding the body on all sides, so that, even if it be separated from either one for a long time, since it preserves the likeness and the imprint it is correctly called an *image*".[26] The power of the sun-mind shapes the lunar substance of soul imprinting her like a seal in wax, which in turn will shape the material-terrestrial body by an enfolding process. The souls on the moon after the second death conserve this medial character, they retain the vestiges of earthly life like dreams and the imprint of the sun-mind. The *visage* is this 'image' of the infinite that has been molded in the finite by the infinite itself, like a wax that has been *finalized* in the literal sense of having been made finite by a mysterious, infinite seal. The *visage* does not have the substance of the infinite but only its image. It is both finite and infinite, both terrestrial and solar, yet its substance belongs in that of the moon, an intermediate body, earth and star. It is an *image* but it cannot be contemplated. It is like a dream.[27] Do we really contemplate in dreams? I do not think so. Yet we 'see' something, for we are in awe.

Dante on the Moon: Visage/Will [Vultus/Voluntas]

As Dante, the visionary astronaut, ascends towards Paradise we find him, at his first celestial station, immersed in the lunar body after traversing the elemental spheres of air and fire. His main concern is the face of the moon. Why does it appear spotted to us? Dante rejects the naturalistic explanation current at the time (which he had once espoused) that the spots on the moon are caused by variations in the density of the lunar body, which would give off or reflect light accordingly. He replaces the theory of the rare and dense portions of the lunar sphere with an elaborate doctrine of the expansion and progressive differentiation of the divine unity of being into the singularities existing in the world. Its Neoplatonic flavor has long been recognized by Dante scholars even though, it has to be noted, medieval cosmology and philosophy were a heterogeneous body of knowledge that defies categorization.[28] The more natural thing to say about it is that the cosmological theory of the spottiness in chiaroscuro of the moon is Dante's visionary concoction from sparse ingredients, culled from the traditional commentaries which formed the motley encyclopedia of Greek, Arabic and Christian sources that, for lack of a better word or laxity of thought, we often still call 'Scholasticism'.

Paradise, Canto 2.[29]

Lo moto e la virtù d'i santi giri,
come dal fabbro l'arte del martello,
da' beati motor convien che spiri;
e 'l ciel cui tanti lumi fanno bello,
de la mente profonda che lui volve
prende l'image e fassene suggello.
E come l'alma dentro a vostra polve
per differenti membra e conformate
a diverse potenze si risolve,
così l'intelligenza sua bontate
multiplicata per le stelle spiega,
girando sé sovra sua unitate.
Virtù diversa fa diversa lega
col prezïoso corpo ch'ella avviva,
nel qual, sì come vita in voi, si lega.
Per la natura lieta onde deriva,
la virtù mista per lo corpo luce
come letizia per pupilla viva.
Da essa vien ciò che da luce a luce
par differente, non da denso e raro;
essa è formal principio che produce,
conforme a sua bontà, lo turbo e 'l
chiaro.

Translation: Allen Mandelbaum.

The force and motion of the holy spheres
must be inspired by the blessed movers,
just as the smith imparts the hammer's art;
and so, from the deep Mind that makes it
wheel, the sphere that many lights adorn
receives that stamp of which it then becomes
the seal.
And as the soul within your dust is shared
by different organs, each most suited to
a different potency, so does that Mind
unfold and multiply its bounty through
the varied heavens, though that Intellect
itself revolves upon its unity.
With the dear body that it quickens and
with which, as life in you, it too is bound,
each different power forms a different com-
pound.
Because of the glad nature of its source,
the power mingled with a sphere shines forth,
as gladness, through the living pupil, shines.
From this, and not from matter rare or dense,
derive the differences from light to light;
this is the forming principle, producing,
conforming with its worth, the dark, the
bright.

According to Dante, the virtue of the divine unity expands from the empyrean through the nested celestial spheres, which work as filters and divisors, thus shaping and influencing the beings that form underneath. Like the soul is articulated in the different parts of the body, so the unity of being is articulated in the universe through the agency of the intermediate celestial bodies, the moon being the last of the divine agents that manoeuvre by receiving less differentiated power from above, and perfect the work of the hammer like the smith, by operating below, thus transforming and further individualizing the power from above. It is a theory of the mystery of existential individuation. Scholasticism never solved the riddles of the principle of individuation, as it struggled to combine the rationalism of Aristotle's metaphysics of being, the pluralistic Greek pantheon, and Judeo-Christian revelation and monotheism. Dante imagines the individual as the final result of a sequence of differential operations (an ontological calculus), the last of which is performed by the moon, which produces the ultimate division after which being emerges into the miracle of individuated existence, into the absolutely 'other', the singularity that has escaped the peaceful unity of undifferentiated divine being.

The chiaroscuro prepares the setting for the existential drama that Dante stages in the lunar sphere. The doctrine of the light and dark spots has generally been dismissed by critics as sterile philosophical discussion interrupting the flow of poetic images that would constitute poetry proper.[30] It has been of interest to historians of medieval philosophy who have tried to make sense of its genetic makeup by looking for antecedents. But in order to appreciate its subtle function in Dante's strategy one has to listen to his dialogue with the blessed spirits that descend on the moon from the empyrean.

The *visage* of a mysteriously beautiful woman speaks to him about her past life, focusing on her breach of a religious vow, and explains the position the 'lunar' spirits have been allotted in the hierarchy of Paradise. The visage of a Florentine woman (Piccarda Donati) reveals her name. She had been a nun of the Order of Saint Clare of Assisi. She was kidnapped from her convent by her brother and forced to marry. Only her *visage* is faintly 'perceptible' by Dante, as if it were a reflection in a mirror, the color of a pearl on a white forehead (Figure I.1). In the heavens above the moon, where Dante will move after his first encounter, spirits will not appear to him in a human form. The manifestation of a *visage* is perhaps the most exquisite moment of realism in the whole Paradise. Its presence is not sensed through ordinary human visual perception, though. The workings of cognition in the heavens is altered by the extraordinary suspension of terrestrial qualities such as gravity that God has granted him. Dante has penetrated the lunar body because his corporeality has been separated from the quality of impenetrability. His eyesight is continually modified by the flashes of luminescence coming from the eyes of his spiritual guide, the blessed apparition of the young woman (Beatrix) whom Dante had loved in his youth. He cannot look into Beatrix's eyes as she is seeing with non-human eyes, namely, with the eyes of God himself. *Visage* is present but not perceptible. The drama of presence/imperceptibility of *visage* is symbolically related to the mysterious chiaroscuro of the lunar spots. It is also the paradox of human free will. The *visage* is an emblem of the frailty and vulnerability of human existence. Piccarda's is a drama of the pliability of will coming into contact with the force of violence. Piccarda is blessed and yet guilty at the same time. That is why she appears to Dante in the lowest sphere of the universe. Why should violence, the immediate cause of her breaching the vow, diminish the merit of the free will, asks Dante?

FIGURE I.1 *Portrait of a woman with a pearl* by Giovanni Agostino da Lodi. From: Dante Alighieri, *La divina commedia. Volume III. Paradiso.* Edited by Corrado Ricci. Milan: Hoepli, 1921, p. 724.[31]

The answer is that Piccarda did not resist the violence, as she should have done, by explicating to the fullest her free will to remain forever faithful to her vow. Human free will is ideally like fire. The nature of fire is such that the flame always returns to moving upwards as soon as external disturbances such as wind cease to influence its natural tendency. But if a flame were to remain inclined or flat after the tempest is over, its nature would not be that of fire. To the perplexed Dante, his spiritual guide explains that Piccarda might have returned to the convent as occasion must surely have arisen. She did not. Her flame was turned downwards by stormy violence forever with her complicity. Her free will became relative—*secundum quid*, according to the Scholastic jargon—as she failed to remain steadfast in the face of brutality. However, she remained faithful to her vow in an absolute sense, that is, in the sense of preserving in her heart, during her forced marriage, a genuine faithfulness to the promise she had made, notwithstanding the adverse circumstances.

Paradise, Canto IV.[32]	Translation: Allen Mandelbaum.
Se vïolenza è quando quel che pate nïente conferisce a quel che sforza, non fuor quest' alme per essa scusate: ché volontà, se non vuol, non s'am- morza, ma fa come natura face in foco, se mille volte vïolenza il torza. Per che, s'ella si piega assai o poco, segue la forza; e così queste fero possendo rifuggir nel santo loco.	If violence means that the one who suffers has not abetted force in any way, then there is no excuse these souls can claim: for will, if it resists, is never spent, but acts as nature acts when fire ascends, though force—a thousand times—tries to compel. So that, when will has yielded much or little, it has abetted force—as these souls did: they could have fled back to their holy shelter.

Some think that *Paradise* is a drama with no psychology, the characters being devoid of human complexity, lacking in depth psychological definition—low resolution types, so to speak, at least by modern narrative standards. This is quite true. But there is in Dante's presentation of Piccarda's dilemma something much deeper than Freudian psychoanalysis, there is the mystery of the coincidence of opposites. For, Piccarda is both a real flame in her heart and a bent flame in the circumstances of life. Both flames animate her destiny. She is a chiaroscuro of the will immersed in the ashen light of a lunar landscape. This is the destiny that the finalizing operation of the spotted moon has crafted for her terrestrial existence as a *unique* incarnation of the same dilemma.[33] The chiaroscuro that every human/animal *visage* reveals. To this chiaroscuro, human/animals can entertain a biosophical relationship that exceeds the epistemic horizon, within which the subject/object violence is played out, and yet which is not purely ethical. The biosophical is beyond the dichotomy epistemic/ethical, for it dissolves such boundaries, and perhaps the conceptual dichotomy is false, the detritus of the imperialism of the subject. Or perhaps the dichotomy is illusory, as the dichotomy finite/infinite also turned out to be. There is indeed such thing *intermediate* between the finite and the infinite, the *greater than any assignable finite quantity*. For, the infinite itself remains a mystery, ungraspable, it is both beyond the reach of thought and away from the ethical impulse. The *visage* is not infinite, it is the harbinger of infinite freedom.

The Physiognomic Imagination

The *visage* is revelation. "From the appearances of the face, without needing experience, almost like a celestial oracle, or a divine art, or the fastest possible divination technique, all of a sudden this art reveals the evil people whom you must avoid, or the honorable ones whom you must embrace".[34] This passage from the *Preface* to Della Porta's *On human physiognomy*, hints at the obscure relation of human language to the unspeakable, namely, the image, the ideograph, the emblem, the geometric diagram, the mystery of the face—human, animal, plant. It invokes the unspeakable element in perception (mostly visual perception) that resists being articulated by the logos, and which strikes us as the revelation of a reality that we cannot otherwise attain. The scientific revolution and its ancillary philosophy, in which Della Porta played a significant though unrecognized role (himself a member of the Academy of the Lynx, the most notorious representative of which was Galileo), demoted the unspeakable to a marginal position, a residue, a discard of the rational activities of the logos manifesting itself in subjectivity, the domineering all pervasive imperialism of the ego thinking its own thinking and falsely deducing its own existence. The analytic form of reasoning that emerged in early modern mathematics defined science as logical argument, namely, deductive argument. Analysis denigrated the image, for the image was irreducible to the strictures of syllogism. I recover Della Porta's project of a universal physiognomy of the face, an art which captures the unspeakable and brings it into emotional contact with the infinity of the human/animal experience. I suggest that a restoration of universal physiognomy is desirable, as reverence for the unspeakable, as responsible habitation in language that reveres the ecological diversity of the human/animal singular existence.

Della Porta's *On Human Physiognomy* is an intriguing amalgam of ancient learning and first hand observation. The definitive edition consists of six books. The theoretical foundations are laid in the first two books, where Della Porta discusses the history and epistemic status of physiognomy, and performs the analytical groundwork that profiles the morphological features of the non-human animal body in correlation with its moral and psychological patterns of behavior. In the third and fourth books, Della Porta delves into the maze of human and non-human bodily, moral, and psychological resemblances, the third book being entirely devoted to the eyes. He discusses details concerning the psychological and moral constitution of both humans and animals, their hidden natures, in relation to their common bodily features that are perceptible to the senses. The fifth book presents a complex typology, it is the book of synthesis following the analytical effort, an ambitious attempt to classify the bewildering diversity of human character from both a moral and a psychological perspective through the construction of abstract characters such as the 'the just human'. Finally, the sixth book offers a naturalistic therapeutic of the character based on a largely pagan conception of the animal body and natural remedies, which looms large, despite precautions, from behind the scenes of Della Porta's preoccupation with Catholic orthodoxy throughout the treatise.[35]

The mystery of animal physiognomy (both human and non human) is the mystery of the presence of animal life in the cosmos. The art of divining the moral character from signs, in this case, from bodily features (the most important of which are the face as a whole, or are located in the face such as the eyes), is the art of glimpsing with the imagination the infinity of the moral character from a finite set of signs provided by visual and other perceptual appearances such as breathing, speech and voice. Let us contemplate Figure I.2 for a moment.

FIGURE I.2 Man and his lion archetype, similar, for instance, in their square foreheads. The picture is from Della Porta 1586, p. 34, but the square face discussion is from Della Porta 2013, p. 123 (note that there are interesting differences in the organization of the descriptive and visual materials among the many editions of Della Porta's treatise on human physiognomy). See Gabrieli 1932, for an orientation in regard to the sundry editions of Della Porta's popular works.

The square forehead of the man, having the right proportion and order, and the mean size of the face overall indicate wisdom, magnanimity, and prudence in so far as such a face resembles the face of the lion.[36] It is a striking claim. First of all, let us note that for Della Porta the lion represents the archetype of man, the leopard the archetype of woman. But, as Della Porta also knows well, signs can be gathered from elsewhere, for example, from the proportion of humors that blend together in the animal body so as to determine its humoral temperament, hence both its moral and psychological complexion. In antiquity a plurality of systems of physiognomy were devised that emphasized different semeiotic systems as bases for divining character. So the question arises as to why non-human *animals* are chosen by Della Porta as archetypal characters, in effect, as the foundation of the *science* of physiognomy.[37] It is a difficult question that, I believe, can be tentatively answered as follows. The animal kingdom offers a variety of patterns of moral and psychological behavior which, according to tradition and lore, are profiled prominently, brought to the fore, so to speak, very often in a single animal or more often a species. This animal displays a character that is well recognizable, almost like a prototype that appears to be perfectly demarcated from the continuum of observable patterns of behavior. Hence animals offer the full spectrum of natural characters that appear to be blended together in human beings and hinted at by their visage. Thus humans are like psychological, or moral, hieroglyphs summarizing many modalities of behavior confused together. The art of physiognomy consists in understanding the whole character, glimpsing its belonging to one type, or figure, or more, by correlating the single bodily features with their corresponding animal modalities in an act of holistic perception or imagination. The natural history of animals forms the encyclopedia of modalities that is the basis for learning physiognomy. The gaze of the physiognomist performs the act of perception and imagination of the bodily features so that the constitution of character is revealed all at once. But this physiognomic activity that cognizes the moral hieroglyph portrayed by a *visage* remains very mysterious. Indeed it must exceed perception, and thus turn into an act of the creative imagination. The moral character can be spoken, in the sense that its abstract figure, or type, can be described in language (as we shall see presently), but physiognomic cognition, resulting in the revelation of character suddenly, from pure perception of the *visage*, attaints to the unspeakable substratum of experience that transcends verbal communication.

Once the analytics of moral modalities to be discerned in non-human animals has been accomplished, Della Porta compiles the inventory of abstract figures that emblematize the typical characters. It is a moral, or psychological typology, a grand synthesis, in which a description of each type is assembled on the basis of a variety of signs collected from the analytics of the archetypal animals. The first and foremost will be the figure of the just man, an original contribution to the science of physiognomy, as Della Porta claims, for antiquity did not describe it. (Note that the figure is a masculine one. Here I must leave aside the problem of the sexism embraced by Della Porta, a piece of cultural baggage which should be dealt with extensively elsewhere for lack of space). Della Porta composes forty-three abstract figures in total, which include, to mention just a few, the simpleton, the evil man, the faithful, the timid, the bold, the magnanimous, the stupid, the tempered, the melancholy, the irascible, the liar, and the hero. The archetypal non-human animal for the figure of the just man is the lion (but also the elephant to a smaller extent). The body shall be well proportioned, the color of the hair dark, the voice of a low tone, resounding and stable, half way through the bass and the treble. The eyes will be large, high, prominent, resplendent and moist, with the circles around the iris equal, that is, the inferior, narrow and black, and the superior of a fiery nature and moist, happy when laughing. Eyelids will be modest, and the square forehead will extend right and left. Traditional lore attributes justice to the lion for it is believed not to cheat and to take revenge proportionally to the injury that it has received.[38] Let us go back to the figure once more to recognize the face of the just man in the archetypal effigy of justice, i.e., in the *visage* of the lion.

The figure of justice is an extraordinary achievement of Della Porta's physiognomy (as he modestly points out in between the lines).[39] It was impossible to synthesize for the Greek mind (according to my reading of Della Porta's understatement) because for Greek philosophy justice was not a virtue to be located easily in a single part of the soul, but rather a concomitant of the whole soul. But for Della Porta the soul receives its moral traits from the temperament and the complexion of the animal body that it inhabits, including the virtue of justice which is classified by the Greeks as a virtue of the intellect. Della Porta's physiognomy is naturalistic through and through, bodily oriented, not materialistic and not reductionist. In the sixth book of the *Physiognomy*, Della Porta investigates the remedies and therapeutic strategies that by balancing the temperament of the humors in the body can restitute integrity to the body and hence restore or perfect the virtue of justice to a certain extent. This last part of the book might be superficially read as an orientation towards a materialistic concept of ethics based on a physiological notion of balance or humoral temperament. But the remedies indicated aim at enhancing health in the holistic sense of well-being. Good foods and medicinal plants take care of temperament, a harmony of the flesh and the mind, which is itself the secret of justice insofar as the tempered body implies justice. Temperament is ineffable, certainly not materialistic. In the harmony of the flesh and mind the distance from modern materialistic medicine is enormous. The human/animal flesh in which the Christian God of Della Porta was incarnated was brought into a harmony with the verbum. Christ embodied the supreme justice.[40]

Justice is diffused throughout the animal body. There is no seat of justice in the human/animal brain. The reductionist concept of a cerebral location of cognitive function that was pompously embraced by modern phrenology and that has been given center stage in contemporary neuroscience (a refined form of nineteenth-century phrenology based on computerized imagistic technology) is suspicious. Justice is infinite, diffused, transcendent, and it transcends the wrong-headed project of psychological localization inaugurated by the Greeks that culminated in neuroscience. Justice emerges unscathed in the human/animal visage announcing that the doorway to infinity is flung open by the call to ethics. It is the virtue of justice that, according to Della Porta, orders and presides over the hierarchy of all virtues by sitting at the top of the pyramid. Philosophy insofar as physiognomy is first and foremost an ethics of the human/animal *visage*.

Della Porta's physiognomic ambition was ethical on a cosmic scale so as to embrace the totality of animal life. It was also trans-humanistic since it connected the non-human animal, the 'irrational', to the domain of human language, where humanism and rationalism inhabit, by brining to bear the creative imagination of the human/animal *visage*. It elaborated the possibility of an experience of the exteriority of the animal *other* that, at the turn of the twenty-first century, resonates more strongly than ever. Physiognomy is an art of the creative imagination that does not separate the human from the non-human. It places the human on the continuum of life, a line in the spectrum of moral and psychological behaviors that are emblematized by non-human animals. Its cosmic ambition reverberates an echo to which we must learn how to respond afresh. At the beginning of the twenty-first century, five hundred years after Della Porta, we confront the destruction of the planetary ecosystem that threatens the very existence of biological life. The scientific worldview that was established in the nineteenth century engulfed the epistemic pluralism of the Renaissance, enforcing a totalitarian epistemology which has colluded with the political totalitarianisms (and the democracies) of the last century. We are experiencing a cosmological catastrophe because we have defiled the *visage* of the world by relegating imagination to the realm of the rationally suspect. But we can learn how to restore the domain of the human/animal *visage*, the legitimacy of the experience that reveres the unspeakable in cognition, for this will be an act of balance that returns us to the epistemic pluralism of the Renaissance. Human habitation of, or incarceration in, language need not negate the unspeakable, but make peace with the many mysterious tenants in the house of being.

By deploying a rehabilitative and therapeutic strategy, the art of physiognomy exposes the humanistic totality, namely, the reification of human relations of violence (both physical and epistemic) towards non-human animals, which has defined Western reason throughout modernity and beyond. Ever since the early modern encounter between human and brute, exemplified in Descartes's concept of the animal-machine, the humanistic totality has grown, like a protean amoeba, by swallowing the brains of the burgeoning natural sciences rooted in the magical Renaissance. Western consciousness has repressed this debilitating cannibalism punctuating the history of modernity. The humanistic totality resists deconstruction and postmodern prosthetics insofar as its metabolism has removed all traces of the originary processes of the natural sciences. Hence the work of the physiognomy of the human/animal, the creative imagination of the human/animal *visage*, proceeds like a series of experiments, where the sublimity of the opus, the transformative moment that defines the subject anew, consists in witnessing the rebirth of the totality at its most originary, embryonic stages of formation. The work itself rekindles the fire of the crucibles in which the non-reductionist scientific naturalism of the Renaissance was transmuted into a totalizing, analytical humanism. Thus the adept in biosophy experiences a detoxification of existence whereby she locates the regions in the cultural landscape that welcome the presence of the animal without incurring any violence. Existence in the process of rehabilitation cannot unfold deductively, like logical reasoning, as deduction itself is the carrier of lethal epistemic violence. Rather it peacefully suffers the animal exteriorities emerging into the experience of *visage*, and allows their apparently fragmented orders of reality to suspend the spatiotemporal individuation that defines the totalizing human. It thus undermines the pursuit of truth as total assimilation into the injustice of the human that was enforced by the corruption of the natural sciences, and which is predicated on the reduction of the animal to an atom of anonymity, subjected to the coercion of deductive reasoning. An ethical philosophy that pursues truth as total assimilation of the other, animal or otherwise, cannot dismantle the humanistic totality, let alone abjure the histrionics of reason on the brink of mute animality, as we shall see in the next section. Fascist humanism will only be dissolved by the consummation of an eschatological event whose meaning cannot be grasped by reason alone. Instead, the human/animal *visage* abrogates truth as total assimilation, indeed it abrogates truth altogether, and resists the urge to spatiotemporal individuation without however negating it but preferring the undifferentiated liquidity of the amniotic currents. It does not enforce a model of reasoning abstracted from the spatiotemporal order of the natural sciences. Rather it explores the orders of existence that constitute human/animality in all its forms. The ethical existence allows an infinity of echoes to reverberate the human/animal presence in the cosmos, the *visage* that refuses to be enslaved into a totalitarianism in the name of reason. It embraces the anarchy of figures, the joyful strife of elemental and humoral motility that defines the orders of possible experience, the sacredness and ineffability of temperament, with respect to which human/animals are individuated as infinite, intrinsically just singularities.

2. The Histrionics of Reason

I return to the theme of useless suffering by imagining myself as an anthropologist from a future time visiting a neuroscience laboratory of the nineteenth century. I am going to meet professor Pierre Flourens, well known for his research on animal cognition, and quoted by Luigi Paolucci. Flourens was the most celebrated neuroscientist of his time in France and one of the foremost European intellectuals and philosophers of science.[41]

I report what I have learned from reflecting on the results, techniques and procedures by which Flourens operates on his research animals in order to study brain, cognition, will and behavior.

My first curiosity is about the type of animals that Flourens uses in his vivisections. As for brain experimentation he chooses especially young animals. Technically it s much easier to remove the cranial bone as it is more tender in younger animals. But what is crucial is that these animals endure the mutilations that are inflicted on them much better so as to serve the purposes of prolonged observation and research more effectively. The denudation of the brain must be done in a circumscribed fashion because these lesions allow the animals to survive longer. Chemical irritants are avoided by Flourens as it is very difficult, as he explains, to moderate the action of these substances. The technique of gradual ablation of the brain tissue allows the experimenter to achieve a rigorous determination of the functions that he is looking for. In other words, an accurate correspondence between localization of regions in the brain and certain specific functions can only be obtained by following the longitudinal progress of the behavioral responses of the animal under the changing circumstances of progressive denudation. The brain, according to Flourens, is the organ by which the animal receives the sensations, executes or determines its movements, and by which it perceives and wills. The research questions that guide Flourens are to all intents and purposes fascinating. Does the faculty of sensation and will reside in the same region of the nervous system as the perceptual property? Does the latter reside in the same area as that of motion? Finally, are thought, sensation, and motion one and the same property? Or are they different properties? Flourens has come to the conclusion that in the nervous system there are three different properties essentially distinct, the first is perceiving/willing, the second is sensation, and the third is motion. They are located in different regions of the nervous system and there are well-defined borders that separate them from one another. Flourens believes that perceiving and willing constitute intelligence. Sensation is the reception and transmission of impressions. Excitability is the property of the nervous system of exercising muscular contraction immediately. The cerebellum is the seat of the property of coordination of the willed movements. In summary, according to Flourens, the intellectual faculties and perception are located in the cerebral lobes, coordination of locomotion is seated in the cerebellum, and the immediate excitation of muscular activity resides in the spinal marrow. Flourens has formed an ambitious vision of the nervous system in vertebrates (to which he researches are confined). The fundamental fact of the brain and the nervous system is that a complete independence separates the functions of the cerebral lobes from those of the cerebellum. From the independence of the organs or regions of the brain he deduces the independence of the functions. Vision and hearing are located in the lobes. But for the other perceptions such as taste and smell the only manner in which he could decide whether an animal would still have the perceptions after ablation of certain layers of the brain tissue or parts of the nervous system was to make sure that the animal could survive as long as possible after the operation. The animals that survive demonstrate that after the ablation of the cerebral lobes, all perceptions, all instincts, and all intellectual faculties are lost. Therefore, they all reside in the lobes, according to him. The second fundamental question that Flourens has investigated is whether these perceptions and faculties reside in different regions of the lobes or in the same one. It turns out, surprisingly, that they reside in the same region, for once one has been lost all the others are also lost. In sum, according to Flourens, they are essentially *one* faculty residing in *one* organ.

Flourens describes the experiments on the cerebellum of pigeons that he carried out in order to determine its role and functions. He narrates the events in a clear and vivid language. I will follow closely his account on my notebooks but omitting some material and detail for brevity.

> I suppressed the cerebellum by removing successive layers in a pigeon. After the removal of the first layers only a general feebleness and a lack of harmony in the movements appeared. At the middle layers a total agitation became manifest. But there was no sign of seizure. The animal moved briskly and in a deregulated manner, but it could understand and will. At the removal of the last layers the animal lost the faculty of flying, jumping, walking, and standing, which had been altered gradually during the previous phase. It was constantly agitated like in a frenzy. For example, it could make out the blow that was going to strike it, it tried to avoid it, it made absurd contortions to protect itself, but to no avail. Eventually, though, volition, sensation, perception, were still present. It was the coordination of the movements that despite the possibility of moving became impossible. I removed the first layers of the cerebellum from another pigeon. I continued my operation. Equilibrium was lost. It could still walk but as if drunk. It would easily fall. At the removal of the last layers all traces of equilibrium, that is, harmony, were gone. What remained was volition. The animal could still want to move but it could not. It is a surprising thing to see the animal that loses gradually its faculty of flying, then the faculty of walking and finally that of

> standing, as it loses its cerebellum. I continued my ablations. The animal lost its faculty of walking totally. As its feet were no longer able to serve its purposes, the animal managed to help itself by leaning on its elbows, its tail and its wings. If I nudged it forward, it rolled on its head. If I pushed it backwards, it rolled on its tail. I removed further layers. The pigeon lost ultimately the faculty of leaning on its elbows, tail and wings. It spun continuously on itself without being able to come to rest in any given position. Exhausted by the fatigue of spinning, the pigeon came to rest in a random position, on its belly or back.[42]

I have been at great pain to relate to the reader my encounter with the spinning pigeon of Flourens. But I wish to say something about my feeling, sure, for what is worth, and frankly. My first reaction when I read the surgically exacting account by Flourens, I have to confess, was a burst of laughter, sonorous laughter, irreverent and devoid of sensitivity. I could not resist laughing with gusto at that image of the spinning pigeon. There was something of the absurdity of the circus in that image, which I had experienced in my childhood. The circus is uncanny, surreal in all respects. It is Fellini's secret and quintessential reservoir of archetypal images. The spinning of jugglers, the spinning of young women in tiny costumes high above flimsy safety networks, or evanescent silhouettes on tight ropes, the spinning of animals playing buffoons with their tormenters. The circus is the archetype of the irreverent joy of spinning. I believe that the torture inflicted by Flourens on his pigeons has something to do with the fascination of the circus, the deadly seriousness of laughter. For, in the circus I came to know realities that exceeded the real. The circus appears like a dream of monsters that look forward to becoming humane. It is the manifestation of the inhumane in the human, and in a flamboyant, obscene, hilarious manner at that. Flourens is a circus character, a clown, a jester, who mesmerized the scientific community of the nineteenth century and still does so, surreptitiously, at the beginning of the twenty-first one. Its buffoonery probes deep into the humanity of the human/animal reader. It makes us feel angry or laugh. I laughed. I have been captive to the archetypal image of the spinning pigeon ever since I bumped into the text. I cannot find peace with this image. I manipulate it in my mind, like a mental model, a toy for torture, for fun, picturing it like a living turbine, fast or slow, a failure of nature to excogitate the biological jet engine, tumbling this way or that. I try to see the torture in the game but I inevitably end up bursting into piles of laughter. I feel victimized.

To exorcise the *Felliniesque* image I conjured a philosophical thought-experiment in the vein of Nietzsche. He thinks that non-human animals forget and live exclusively in the present, or thereabout. That is why nature has preserved them from the corroding power of memory and history, which makes humans decay, become sick, and eventually perish. If you ask the animals a question, they will answer but before they speak they forget what they want to say to you. So, I asked myself what the decerebrate pigeon might experience while spinning. Does the world appear to it turning upside down? Or has the memory of the world been lost owing to the ablation of the cerebellum? Is a pigeon-world still possible after the cerebellum's function of coordination has been obliterated? If the world is a totality of coordination, thoughts and sensory-motor activities, would the world survive the operation, or would it become an impossibility, or a half-world? Does the pigeon experience the violence, the useless suffering, the pain caused by the histrionics of reason? Or does it forget all about the world? Kafka's hero turned into an insect, healthy for a while, feeling relief when moving like an insect. I turn into a spinning pigeon, decerebrate, incorporating the minutiae of the non-human character (I am following Della Porta) from the mutilated body of the pigeon. No relief. No world. Pure spinning, like the acrobat in the circus.

Unlike the Western philosophical notion of truth as total assimilation, a wanton preying on the amiable weaknesses of its face-less objects or non-human subjects, the circus brings to the fore another type of truth, uncanny, supernatural, uncomfortable, but more in tune with the exigencies of a future human/animal ethics. It is that of truth as risk, even mortal risk. Truth as trial, the putting to the test of one's assuredness and being, one's lifestyle, an ordeal-like notion of truth, the uncertain outcome of flying over the flimsy safety networks, of risking the wild animals getting sick and tired and turning on you. It is a Franciscan notion of truth indeed, in a most genuine sense. Francis of Assisi was a supreme buffoon, the jester of God and of truth. It is the denudation of one's being by renouncing one's possessions, poverty for the sake of risk, of living at the mercy of the totally unpredictable other. I have learned from the histrionics of reason that there is no truth as long as the human animal remains safe, cozy in their cocoon, reason, never bursting out into the world of surviving skills, of righteous anger against injustice, never attaining the state of uncertainty about one's life, which the non-human animal daily knows perfectly well. Cultural conditioning made me feel unworthy when I burst into laughter in Flourens's laboratory interviewing the master. Yet laughter was salutary, as it forced me to rethink prejudice and truth, brutally.

3. Gryllus on Animal Parrhesia

Paolucci may have been the first European naturalist to evoke Plutarch's dramatic defense of animal superiority (in his infamous *Beast are rational,* also known as *Gryllus*, from the name of the main character, a man turned into a pig by the sorceress Circe, whom she allows to harangue Ulysses on animal exceptionality) in a scientific work devoted to birdsong. The evolutionary framework of ideas in which Paolucci self-consciously places himself harmonizes with Plutarch's surreal achievement. But at the same time it creates a tension between Paolucci's recognition of the baffling phenomenon of animal continuity with humanity and the urgency of the dilemma posed by the demarcation between human and non-human animality. Paolucci addresses the conundrum in his concept of birdsong as animal music, and indeed one might go so far as to say that his approach to birdsong as music pays an unconscious tribute to Circe's harangue. Here I will emphasize the sharp contrast between the Greco-Hellenistic mind of Gryllus and the denigration of animality that has characterized late medieval and early modern Christian ideologies and their bastard daughter, *scientia*, science (or, to put it more philosophically, the dialectical transformation of Christian scholasticism into modern science). I will compare and contrast Plutarch's Gryllus with early modern Christian apologetic literature, in which animals are ruined, emasculated, ablated from the realm of the thinking substance, and portrayed as bestial inanity, as blindly aggressive violence. I will raise questions concerning the mystery of the non-human animal voice.

Jacques-Bénigne Bossuet wrote extensively about human and non-human animals.[43] The *Catholic Encyclopedia* portrays him a follows.

> Such was the true Bossuet. In his life we cannot always find the daring of his eloquence, nor in his conduct the audacity of his reasoning. This great dominator of the ideas—one might even say of the intelligences—of his time suffered himself to be dominated more than once by the thoroughly human dread of being disagreeable and, above all, of giving offence. "He has no joints", he himself said of one of the gentlemen of Port Royal who was somewhat lacking flexibility; to which the individual in question retorted: "And as for him, you may tell him that he has no bones!" The strong, concise mot sums up all the reproaches that can be made against this great memory. Had his strength of character and his apostolic vigor equaled the force of his genius, he would have been a St. Augustine. Falling short of St. Augustine, a Catholic and a Frenchman may be permitted to believe that it is still something rare, something exalted among men to have been merely Jacques Bénigne Bossuet.[44]

How ironic that lack of vigor and force must have failed Jacques Bénigne, for, as we shall see, in his bombastic rhetoric about human superiority he insisted on non-human animals's having this splendid vigor though unpremeditated, unreasoned, unwilled, and hence worthless.

Animals are like music instruments, says Jacques Bénigne.[45] We use them by tuning them to our own purposes. They are like strings that need tension to play in tune and in the right tuning. We condition a dog, or *touch* him, as we do with the strings of a lute, until it sounds perfectly in tune. A taming *toccata* (from touch) is what we artfully do with animals in order that they might play the services that we require of them. But with humans it is different. We teach them. This means that we speak to them until we have a feeling that they have penetrated into our thought. There is no question. Animals might never respond to human speech and penetrate into our thoughts. They are unteachable yet playable. Furthermore, contrary to appearances, they do not learn from each other. When nightingales sing to one another's tune, in a concerto, or parrots repeat the words they have heard, they are but strings sounding in unison. Like the strings of music instruments that are tuned on the same note they respond to the same note that is being played by sympathetic resonance. They echo one another. But when humans learn how to sing, they artfully learn the techniques of music, they do not respond to the same note by resonance but learn how to transpose a melody (some at least do), how to sing a cadence, as long as they have a modicum of natural predisposition, or shall we say, the gift of perfect pitch. It is this ability to articulate pitch, this flexibility of learning the rules of music, that demarcates human singing from bestial vocalization. Humans, unlike animals, really have no 'musical bones', very much like Jacques Bénigne, they sound peacefully to any tune because of the human-all-too-human dread of being disagreeable. They make peace with themselves by going acoustically 'spineless', by singing *invertebrate*, like jellyfish, worm, and octopus.

But it is freedom that demarcates the human from the beast.[46] It is freedom that allows the human soul to strenuously oppose passions, ultimately with such force that it can neutralize them. Where passions dominate unrestrained, on the other hand, the body and its movements hold sway. Hence reason deserts these omnipotent bodies. The great power of the will on the body consists in this prodigious effect, namely, that man is a master of his body, to such extent that he can sacrifice his body for a supremely superior good that he wants to pursue. To throw oneself in the fray, as if on the impulse of a blind impetuosity, as animals do, does not denote anything superior to the body. In fact a glass will be shattered to pieces by falling downwards simply because of its own weight. But to make a determination to die for a reason, self-consciously, notwithstanding all the oppositions of the body that resists this deliberation, denotes a superior principle to the body, and the only animal that possesses this principle is man (whether, then, according to the article of the Catholic Encyclopedia, Jacques Bénigne belongs in the genus homo becomes a matter of fine discrimination—did they take into account his views on human and animal behavior? One wonders.). Yet they reason!, seems to exclaim Jacques Bénigne, in a moment of vertebrate inspiration, but no, he hastens to assuage us. Be at peace, dear reader, they don't. It was God who wanted them to display reasonable patterns of behavior. God painted on the surface of the earth, like Michelangelo on the vault of the Sistine Chapel, a divine comedy of animals that simulate human reason, giving us images, or planar models of refinement, reasoning, virtue, providence, care for the offspring, in sum a puppet theater for human edification. Every animal has a role in this painted comedy, and all that the beast demonstrates is the admirable art and science of its monotheistic author, the divine architect, or painter.

The pagan Della Porta had an unorthodox view of the divine animal comedy. To him the beast had a genuine *visage*, indeed the bestial was the human in disguise, appearing purified, archetypal, essential, nude, emblematic, a harmony of animal modes of existence, pristine and unheard of, a concerto of virtues and vices that humans amalgamate and exaggerate, or distort, sometimes making themselves just or only ridiculous. For the Christian Bossuet, the face of the beast is the caricature of *visage*, the fixated, bi-dimensional, flattened, blind omnipotent body that perishes in the furor of its unswerving instincts rather than calculate the fine casuistry of the higher goods for which it may or may not deliberate to sacrifice its bi-dimensional, *world-poor* existence. Perhaps this Bossuetian bi-dimensionality of the animal comedy is what really illuminates Heidegger's occult metaphor of the world-poverty of the animal. No surprise, if one recalls (how many enthusiastic Heideggerians will?) the staunch Catholicism of the great existentialist (and the scholastic roots of his existentialist philosophy buried in its doctoral dissertation).

In both Della Porta and Bossuet we catch a glimpse of the epoch-making distance between the ancient pagan mind, modeled on the pluralistic Greek pantheon, and modern Christian consciousness. Bossuet's Catholic apologetic profiles the complete reversal of the value system propping up the philosophical categories of Greco-Roman antiquity, an upheaval which was brought about by modern European civilization and the scientific revolution of the seventeenth century. Plutarch's Gryllus offers us a prophetic response to Bossuet's down casting of the beast. The speaking pig is not an animal instrument grunting to the tune of Ulysses's humanist philosophy. He sings a song that we moderns can hardly perceive as musical. The metaphor should be taken deadly seriously. He sings in an idiom that modern humanity may not be acquainted with, not for lack of talent or musical predisposition, but rather for loss of philosophical categories. Plutarch's masterful essay has been studied and admired (or ridiculed) by generations of scholars.[47] But I will summon the ghost of Michel Foucault, ruminating on Greek *parrhesia*, in order to cast light on Gryllus's swinish assault on Ulysses.[48]

"Etymologically, 'parrhesiazesthai' means 'to say everything'—from 'pan' (everything) and 'rhema' (that which is said). The one who uses parrhesia, the parrhesiastes, is someone who says everything he has in mind: he does not hide anything, but opens his heart and mind completely to other people through his discourse".[49] Gryllus is to all intents and purposes an animal parrhesiastes.

> The parrhesiastes is not only sincere and says what is his opinion, but his opinion is also the truth. [...] If there is a kind of 'proof' of the sincerity of the parrhesiastes, it is his courage. The fact that a speaker says something dangerous—different from what the majority believes—is a strong indication that he is a parrhesiastes. [...] Someone is said to use parrhesia and merits consideration as a parrhesiastes only if there is a risk or danger for him or her in telling the truth.[50]

Gryllus meets these criteria. He demolishes Ulysses's conception of human superiority, addressing him in bold assertions that ensnare the hero of the Trojan war in a dreamy vortex and disorient him gradually as he seems to lose his initial arrogance. But Gryllus runs many risks. We do not know whether Circe might have acceded to Ulysses's initial request that she should return the Greek men she had magically turned into beasts to their former human form and dignity. Hence we do not know what might have happened to poor Gryllus in this event. Moreover, no sooner had he started his argument than Ulysses began hurling abuse and scorn at him, predictably. The pig knew. He took his risks. Yet Plutarch does not say that Gryllus is a parrhesiastes. On the contrary, he makes an even bolder, disconcerting move centered around the word *parrhesia*, which is used by Gryllus at 987F (following the standard notation for Plutarch's corpus).[51] He has Gryllus say: "These facts make it perfectly obvious that bravery is an innate characteristic of beasts, while in human beings an independent spirit [parrhesia] is actually contrary to nature".[52] The pig claims that not only is parrhesia not to be found in humans, a claim that would have simply weakened the notion that parrhesia is frequent in human discourse and social behavior, but that parrhesia in humans is contrary to nature. It is human nature that cannot be parrhesia, for parrhesia is a negation of the human. There is no such thing as open heart, frank, truthful speech in humans. There is no taking risks, courage, because parrhesia is contrary to the human constitution. Furthermore, according to Foucault, conscious use of parrhesia implies that the statements made by the speaker carry a severe criticism of the interlocutor. Only thus is the speaker playing the parrhesiastes. But the thoroughly human dread of being disagreeable, says Gryllus, does not pertain to humans. How can we humans fulfill the role of the parrhesiastes, then? Never. Let us see how Gryllus perverts the reasoning by Bossuet that animal valor is inferior to human freedom by upsetting the applecart of the French apologist and playing the only honest, purest, superb parrhesiastes that we might possibly find in ancient and modern literature.

The punch-line by Gryllus that I started from comes at the end of a passage where he has exposed some basic facts about animal behavior that Ulysses apparently ignored.

> Beasts never beg or sue for pity or acknowledge defeat: lion is never slave to lion, or horse to horse through cowardice, as man is to man when he unprotestingly accepts the name whose root is cowardice. And when men have subdued beasts by snares and tricks, such of them as are full grown refuse food and endure the pangs of thirst until they induce and embrace death in place of slavery. But nestling and cubs, which by reason of age are tender and docile, are offered many beguiling allurements and enticements that acts as drugs. These give them a taste for unnatural pleasures and modes of life, and in time make them spiritless to the point where they accept and submit to their so-called 'taming', which is really an emasculation of their fighting spirit.[53]

It is only under the effect of drugs that captive younglings can be tamed. Drugs are the means of control that humans use to turn animality into slavery. Imagine the most lethal drugs in our society such as television, tobacco, and alcohol. Imagine youth and their revolutionary spirits. Young people are turned into a larval state of existence by television and alcohol conditioning. Alcoholic intoxication is a problem that has reached dramatic proportions in colleges throughout the United States. Should education not teach how to be a parrhesiastes? Yet youths are tamed by adults for profit and for purposes of social control. Youth is animality. Society knows that. Hence we extol the intelligent Ulysses above all and promote drugs that cure the 'disease' of youth.

Before his punch line, Gryllus the parrhesiastes put forward a vision for Ulysses to contemplate. It is not an argument, as the text itself implies and the role of the situation of the parrhesiastes requires.[54] But there is a problem with modern renditions of Plutarch's language. Greek is a language innervated by mental categories far removed from modern philosophical concepts. It is a language of visionary tales more than arguments, despite the fact that modern, especially analytic philosophy (in the wake of Christian scholasticism) has perverted the reading of classical philosophical literature. William Helmbold, the valiant editor and translator of the Loeb edition of Plutarch's *Gryllus*, presents Gryllus's vision as an overarching argument. The Puritan rhetoric of English that is sold to millions of colonized students and scientists around the world by footmen teachers and lackey peer-reviewers, namely, the dogma of direct, active speech, aggressive action, as a celebration of the active verb brutalizing the direct object, the colonialist raping the native, cannot stand the imagistic force of the repeated, subordinate unfolding of the Greek participle that strangles the patience of the modern reader, accustomed to the easy-going syllogisms of Christian scholasticism, and by which the swinish parrhesiastes has Ulysses admire the horror of the moving images of animals combating to the bitter end and dying for the sake of undiluted valor. Gryllus's visionary grunting is tamed by the positivist, Darwinian Helmbold as follows, though philologically very honestly.[55]

> Wild beasts, however, you will observe, are guileless and artless in their struggles, whether against one another or against you [i.e., humans], and conduct their battles with unmistakably naked courage under the impulse of genuine valor. No edict summons them, nor do they fear a writ of desertion. No, it is their nature to flee subjection; with a stout heart they maintain an indomitable spirit to the very end. Nor are they conquered even when physically overpowered; they never give in their hearts, even while perishing in the fray. In many cases, when beasts are dying, their valor withdraws together with the fighting spirit to some point where it is concentrated in one member and resists the slayer with convulsive movements and fierce anger until, like a fire, it is completely extinguished and departs.[56]

The gruesome scene rendered by Helmbold as an objective presentation of facts about zoology in the third person, the language of empirical science, does not capture the vividness of the original vision. I must beg the pardon of the reader and present him with a comparative rendition into Italian, to my mind one of the best in a modern European vernacular. It comes from a Renaissance translator that more successfully sustained the chain of participial unfoldment strategically employed by Gryllus in order to mesmerize the incredulous Odysseus in this crucial text. It may have been this text that Paolucci read, as it was edited in the first half of the nineteenth century, even though he quoted the customary title of Plutarch's essay in the original Greek.

> Tu vedi i combattimenti delle fiere fra di loro e con gli uomini, come sieno senz'inganno e senz'arte, e come si difendano e facciano vendetta con palese e nudo ardire e con verace forza, e come non per comandamento di legge, nè per temenza di pena da darsi a chi abbandona il campo, ma per natura fuggano l'esser vinti, e soffriscano gli ultimi mali per mantenersi invitti. Perchè ancorchè le membra più deboli rimangano vinte, nondimeno non cedono, nè mancano d'animo, anzi aman meglio di morir combattendo. Ed in molti avviene che esalando l'anima, la generosità del cuore e la forza si ritira e s'ammassa in qualche membro, e resiste all'uccisore e salta e si sdegna fino a che in guisa di fuoco non è interamente spenta e svanita.[57]

In this magnificent translation, the Renaissance writer managed to keep the dramatic vision that Gryllus puts before Odysseus's eyes fully alive. You see, you see, you see…, implies Gryllus at every turn of the text, an implication that miraculously hinges on the initial exclamatory statement at the opening, and which Helmbold could only suffocate in an innocuous parenthetic position. But the seal of approval comes at the end where the Italian rendition correctly acquires the strong moral connotation that Gryllus the parrhesiastes insists on throughout the essay. The energy withdrawn to that indivisible point in a bodily member, the last point still living in the omnipotent body, the soul which is still fighting, the focus of action and of indomitable suffering, resists by leaping upwards and forwards like fire and *turning indignant*, for it does not want to morally suffer the insults of the attacker.

Gryllus speaks the truth with an open heart. He does not argue with Odysseus because the parrhesiastes is supposed to know the truth, and the only proof of this state of affairs is in the fact that he is in a position to take risks by exposing himself to the potential retaliation of the offended interlocutor.[58] Truth is risk taking, the "human dread of being disagreeable and, above all, of giving offence". But this parrhesia is contrary to human nature, as Gryllus warned us. Truth is accessible only in the world of the non-human animal where life processes are lived-through to the bitter end. It may be that Odysseus's and our obsession with truth and rationality are a red herring. It is conceivable that the idea of truth that the scientific revolution and most traditions in modern philosophy have incorporated into their ideologies, i.e., truth as mirror-like reflection of the states of affairs, facts, obtaining in the world, an idea which remains fastened in the heart and mind of every scientist despite the tirades of philosophers of science of all persuasions, is simply a delusion. Knowing how the world is made without putting the human knower at risk, even mortal risk, is a contradiction in terms. They, the animal others, know better though they do not speak. We are the poor-in-world despite language.

The Voice of Gryllus

Perhaps one way to shift the perspective of violence and aggression from which European science and philosophy (especially analytic philosophy) have regarded the presence in the world of the non-human animal is to consider the phenomenon of the their voices. Plutarch has Gryllus speak to Odysseus in Greek as this is the privileged ontological realm that Odysseus inhabits. Sound is an originary phenomenon in the history of life in the universe. It is perhaps the most archaic manifestation of an illusory dissociation between psychic and material events that became the prelude to the descent of consciousness into the animal kingdom. In shamanic cultures, regardless of whether shamans make use of hallucinogens to enhance their ecstasies, sound and voice are important means of controlling the outcome of an ecstatic experience which tends to dissolve the boundaries between ego-consciousness and the external world.[59] Recovering the archaism of the experience of sound, its harkening back to a shamanic or non-human animal world, is essential for welcoming the strangeness of the non-human animal voice into the life-world of speaking humans. This illusion was encapsulated in the mechanistic dogma on which the entire edifice of modern science has been built. Thus a prerequisite for overturning the dogma that has enslaved perception to the logic of subjectivity (i.e., the argumentative structure of ego-consciousness) and its alleged material correlate, the neurophysiology of the brain, is the recovery of premodern and even archaic modes of life. The latter reflected on the human experience of the world by means of categories and patterns of behavior that did not privilege subjectivity and the brain, but rather emphasized the transcendental aspects of the senses as pathways to accessing the realms of the other regardless of the forms—human, non-human animal, sacred, demonic, epileptic, divinatory, augural, prophetic, sexual—in which the other manifests itself. We have been mesmerized by the emergence of articulation in the human vocal apparatus. Articulation inflects the sound wave, so to speak, that originates in the flute-like apparatus of the trachea and lungs in which a column of air is made to vibrate. The complex mechanisms of muscles and other tissues that govern articulation in the lungs, the larynx, the tongue and the oral cavity of the mouth allow humans to sharply demarcate the acoustic space of speech in terms of vowel-like sounds and consonant-like sounds. There is a bewildering diversity among human languages in the range and number of phonemes that are available to speakers, and the distinction between vowel and consonant is tricky enough to suggest caution. However, in languages such as classical Greek, Latin and the modern European vernaculars, in whose acoustic spaces the categories of modern science, esthetics, and more generally philosophy have evolved, the categorization of vowel-like sound and consonant-like sound is well defined enough, at least from a psychoacoustic viewpoint. Modern digital technologies available to anybody armed with a laptop computer and free software easily allow the visualization of the sound phenomena of phonetic speech. It is instructive to observe the continuity of the frequency spectrum of speech sound waves in which an alternation occurs of vowels and consonants. As vowels tend to last longer than consonants and are sustained by the flow of air in the vocal apparatus, their frequency energies tend to occupy the lower or middle portions of the scale of frequencies in normal speech situations and manifest a more discrete distribution of harmonic content, whereas the consonants spread more over larger portion or even the entire range of the scale of frequencies (with significant differences, however, the details of which are not relevant to my discussion). What is important, I think, is that consonants appear with a less discrete distribution of harmonic content than vowels, that is, their spectral composition is relatively continuous, certainly much more continuous than vowels.

As Greek music theorists had noted since Aristoxenus, music tended to sound (to them) intervallic, that is, the voice was perceived like moving in a voice-space by discrete jumps and remaining stable at certain levels for certain periods of time. In contrast, speech was perceived as moving continuously in the space of the voice. In modern psychoacoustic parlance, music-like sound waves would be described, based on the visualization of the spectra of frequencies, as concentrations of energy on the vertical scale of frequency remaining constant over a duration of time along the horizontal temporal scale. Spectra of frequencies are three-dimensional diagrams that typically use colors to add the information of the evolving energy content over time. Speech-like sound waves would manifest a continuous evolution of the concentrations of energies along the temporal dimension, with little or no zone of constant stability. Thus, we have two impressive phenomena in the human perception of speech-like sound that are important psychoacoustically. First, the dichotomy intervallic-continuous, which was noted by the Greeks as characterizing their perception of music as opposed to speech, and second, the dichotomy vowel-consonant, whose fine grain psychoacoustic structure has become more and more evident in recent scientific and technological developments. There is therefore a striking dualism in the psychoacoustic manifestation of sound phenomena. It is true that contemporary classical music (I use this arbitrary label to roughly indicate the evolution of modern musical esthetics since the revolution of atonality that ushered in the upheavals of electronics, concrete music, serialism, spectralism, noise, minimalism, etc.) has moved away from the Greek conception of music as having a marked intervallic character. However, as the easy visualization of sound waves and dynamic spectra of frequencies (spectrograms) became available, it also became clear that the presence of intervallic concentrations of energy along the frequency scale as opposed to continuous temporal evolution (the Greek intuition) is a rather general phenomenon that manifests itself under the most diverse circumstances such as speech and song regardless of style and esthetic tendencies. In other words, it seems to me that this striking dualism in our perception of sound phenomena cannot be regarded as a neutral manifestation of acoustic patterns that have had no influence on our conceptual frameworks. Thus, from this point of view, the appearance of philosophical and modern scientific categories in Western civilization must be related to the hidden dualism, phonetic and psychoacoustic, inherent in classical languages and in most modern European vernaculars.

The sprightly presence of the consonants as opposed to the insistence of vowels in our perception of speech cannot be dissociated from the elusive demarcation of subjectivity and objectivity. The originary dualism of verb-phrase and noun-object also cannot be seen in syntactic isolation but must be understood in the context of the patterns of alternation of energy concentrations inherent in sound spectra. Possibly it can be related also to the physiologically basic alternation of breathing in and breathing out. It is conceivable, therefore, that the emergence of the dualism subject of action vs. object, or active verb/direct object, in philosophical speech has its roots in these patterns of phonetic alternation typical of such languages as Greek and Latin. Dualisms are all too evidently a deficiency, a poverty if not a disease, affecting the systems of categories that have formed the backbone of Western rationalism and science. They crop up everywhere, in many guises, but obscure rather than clarify the underlying problems. It will be sufficient to mention, for example, the inexhaustible discussions about such correlative notions as cause and effect, or the ego, mind and individual identity. Since Aristotle's theorizing and the Christian Scholastics that elaborated on his views (and setting aside more recent attempts in analytically oriented philosophies, which do not seem to make any progress as well, even on their own terms), they have become neither clearer nor more explicative, or indeed more fruitful. Rather they seem utterly inadequate when we approach mysteries such as the languages of non-human species and their styles of communication. It may well be that communication from humans towards other species and vice versa will not be possible as long as human language is trapped in these diseased rationalistic dualisms. Rationalism tends to divide and conquer, and it is no surprise that Western rationalism has resulted in the most ferocious colonialisms and genocides ever occurring on a planetary scale. Implicitly, as I have suggested, Gryllus warns Ulysses and all of us that only in the animal kingdom is the most genuine parrhesia to be found. But it is not a rationalistic way of life.

To balance the above criticisms, it behooves me to make a positive suggestion as to the possibilities that non-violent modes of communication based on non-dualistic acoustic patterns might open up in the moral equilibrium of biological life on earth and in larger-scale ecologies among diverse animal species. The idea of resonance offers an interesting, alternative pattern that seems to transcend the divisive, disruptive cutting of the acoustic space into consonant-like and vowel-like sounds. It will become possible to speculate on the philosophical categories and modes of life and novel ethics that will emerge in the wake of communicative styles based on resonance. Gryllus's challenge to humans would be met if we might develop such styles of non-violent communication.

Dualisms such as that of subject/object imply a directionality from the subject to the object. A line of communication is open when one being turns into a signal transmitter and the other into a signal receiver. It is unidirectional, as it implies that something passes from the sender to the receiver, a letter, an email, a smoke signal, or a bullet that is going to kill. The roles can be swapped but not the intrinsic unidirectionality that is reestablished at each turn of the communication process. On the other hand, resonance is a feedback phenomenon. To illustrate it simply, I suggest an electronic analogy we might be familiar with. When we tune an FM radio (or used to before the digital era came about) to the exact FM frequency of our preferred local station, we have in effect established an electromagnetic resonance between the electronic circuits of the radio transmitter and those of our receiver. In acoustic resonance, something precisely like that produces an analogous phenomenon of tuning that amplifies the communication. Bossuet's example of the strings that respond to sounds to which they are tuned is correct. Strings can respond by sympathetic resonance, which is a colloquial way of speaking that describes the phenomenon of acoustic resonance. In resonance the frequencies of the participants in the process are set close to one another so that a feedback of energy amplifies the signal to adequate levels for both participants (sometimes too much as it happens in certain resonances such as the nasty microphone feedbacks on stages). In resonance a tuning emerges that is not dualistic and not unidirectional. The accord between the participants happens in such a way that amplification makes the voice audible for the participants. Energy is shared between the resonant beings, and it takes at least two beings to produce a resonance, so that there is no prevarication or violence done by any of the *resonants* towards each other. Resonance transcends egoistic individuation. In this style of communication both of us, you and I, share a common tuning. And more than two can participate. We amplify one another in the presence of a common environment, replete with possibilities of exchange and stimuli, and we apply a mutual endeavor to communicate with one another peacefully. Another way of saying resonance is 'tuning'. Beings are naturally tuned in many respects. What I am suggesting in the wake of Gryllus is nothing really new. Tuning is common in all cultures that have music and sound, which means not only human but also non-human animal cultures. We know that hearing in many species, definitely in mammals, is based on tuning. There are myriads receptors in the human cochlea, in which somehow tuning processes transform acoustic energy into patterns of attunement to the world. It is hard to believe, it is puzzling that we Western humans have gone down the road of the dualisms of the articulated voice, the diseased consonants and vowels, and produced such destructive behaviors that have imperiled the existence of biological life on earth.

I wonder if Gryllus is perhaps unconsciously trying to communicate with Odysseus by way of resonance through his insistence on animal parrhesia. Plutarch is a spectacular writer. He is capable of condensing in a short essay like this a panoply of kaleidoscopic insights into such mysteries, and he may have hit on the fact of resonance without need for theorizing, or even better than by way of theory (which is stuck in, or impeded by, language). It is not easy to categorize Gryllus's style, as scholars have noted, by attributing its 'vagaries' to this or that genre. But if resonance is at work here, we should expect this. Gryllus parades images before Odysseus, even gruesome and horrific ones, as I have emphasized. He is not really practicing the logic of demonstrative syllogisms. And images do have a strange, 'resonant' power in the human imagination, a fact that premodern philosophers such as Della Porta understood quite clearly. Resonances occur also in visual phenomena, not only in psychoacoustics. Note that resonance is a general pattern showing up wherever there are vibratory processes, be they electromagnetic or optical or mechanical or other ones. We might even say that in the scientific understanding of the world it became especially important following the ascendance of the notions of field and the quantum in twentieth-century physics. Resonance is systemic. It is not only between you and me. It is a systematic manifestation of biological life on the ecological scale. One might even suggest, on the cosmic scale. Hence it transcends the isolated individuals.

Resonance is transformative in a gentle way, like the moon that operates the gentle death (the second one) by separation of the intellect from the souls that have left the terrestrial body. It is not an assimilation of the other. It is not autocratic and autonomous. It is not violent. While Odysseus starts off by hurling abuse at Gryllus and making insinuations on his allegedly original swinish predisposition, the gentle pig looks for resonant spots that might open up communication. Whatever rhetoric and aggressiveness are left in his speech, they are a residue, almost an automatism occasioned by Odysseus's arrogant remarks. After all he still speaks philosophical Greek, willy-nilly. The hero would not understand *piggish*. Odysseus goes through a fantastic transformation in the course of the resonance. I might even say 'transfiguration'. Yes, there is an element of that in the magical process. It is a transfiguration that changes the aggressive nature of the solar human into a more subdued, moon-like, feminine presence poised to be welcomed somewhere along the continuum of animal sentience. He softens quite considerably. His tone has changed. We might speculate that his comments, perhaps not all of them registered by the author, have become more pensive and sparse, and by the time the game is over the reader has lost confidence in him. The arrogant intellect of the king of deceits has been separated by Gryllus's parrhesia. The concluding question about the concept of God, which Ulysses asks Gryllus as a last-ditch, desperate attempt at defending the indefensible, human superiority writ-large in their motley religions and belligerent pantheons, is made to evaporate like thin air by the pig's en passant recollection of the descendance of the hero from a notorious atheist. After which moment, silence looms large. Resonance fades off into the music of emptiness.

Papparotti and Resonance

We cannot do an experiment with magic pigs. But I invite the reader to watch an extraordinary video on the internet. A parrot named 'Eric' sings along with the tenor Luciano Pavarotti a celebrated Italian song (written in the 1980s by the avant-garde songwriter Lucio Dalla, and entitled 'Caruso' in honor of the iconic Italian tenor of the early twentieth-century).[60] In fact Eric is not imitating Pavarotti but really duetting with the famous opera singer. I nicknamed the superstar parrot 'Papparotti' (from the Italian 'pappagallo', meaning 'parrot'). It is an extraordinary performance. Parrots have laryngeal structures somewhat similar to those of humans hence they may be able to reproduce sounds that are quite similar to those of human origin. In Figure I.3 I have reproduced a diagram of the spectrogram of a short phrase at the beginning of the video, where the frequencies of Eric and Pavarotti are partially overlapped as Eric is starting to sing his line shortly before Pavarotti. I added a circle to indicate the moment when Pavarotti in turn begins to sing the same musical phrase that Eric has just started. The lowest thick lines (very thick and irregular!) represent the approximate values of the frequencies. I have not reported the vertical scale of frequency as it is irrelevant for my considerations. In this picture pitch is represented by the lowest thick lines, the lines above are the so-called harmonics, while time evolves from left to right on the horizontal axis. The reason why the lines are not very thin but instead very thick and fat is that the spectrogram cannot zoom in like a microscope beyond a certain limit, for there are mathematical constraints on the resolution of these spectral analyses (which however do not seriously affect the reasoning that we can extrapolate based on them). The first three thick, bottommost lines in horizontal succession from the left, beginning where the word 'Eric' is placed in the diagram, correspond to the syllables 'Ti', 'vo', 'glio', from the phrase 'Ti voglio', which Eric scans with surgical precision, and indeed they appear clearly delineated and well separated in the spectrogram. The pitch is raised two times, from the first to the second syllable, and from the second to the third. The circle indicating where Pavarotti starts to sing is placed in the region where there is an almost complete lack of frequencies because that is the moment when the text continues with the word 'bene', and Eric takes a little while to pass from 'glio' to 'be'. The consonant 'b' on the other hand has no clearly defined pitch and therefore no discrete harmonic content appears in this place (there is some in the original spectrogram but to simplify the reading I have added a threshold and made the diagram purposely less accurate for the sake of clarity). It is a coincidence that at this moment Pavarotti starts the same phrase Eric initiated three syllables earlier (is it?). After that moment they continue together in unison on the syllable 'be' for some time (starting where in the picture you can see the label Eric+Pavarotti and ending roughly where the label ends, at which moment Eric goes silent and Pavarotti is solo until the end of the spectrogram). While they hit the same pitches their voices do not have the same harmonic contents. The region where Eric's harmonic content is stronger is higher that that of Pavarotti's. As the reader can verify the timbres are certainly different (though not that much, to my ears).

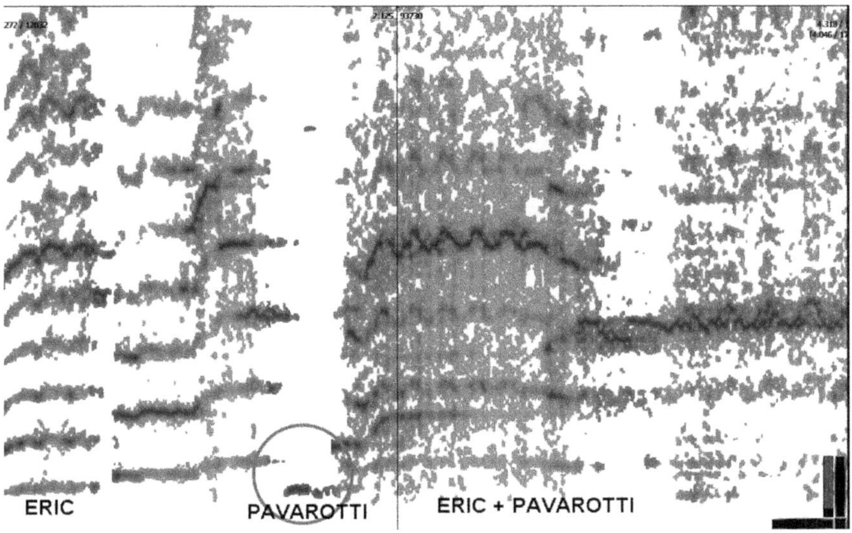

FIGURE I.3 Spectrogram of the initial segment of the song starting at 28 seconds from the beginning of the online video. The circle indicates the moment when Pavarotti begins to sing the same phrase that Eric started. The lowest thick lines represent frequencies. So pitch is represented in this diagram by the lowest lines while the above lines are the harmonics. (The bars in the bottom right-hand side corner are an artifact of the diagram). The spectrogram has been produced with *Sonic Visualizer*, a powerful free software available on the internet (http://www.sonicvisualiser.org/).

I recommend that the reader watch the entire video as some reflections I will make are based on the patterns of behavior that continue to the end of the performance. Unlike many examples of animal-show parrots that imitate human voices, here Eric appears to all intents and purposes to really 'be in the performance'. Not only are its lines generally in harmony or unison with the tenor but the timing, the anticipations and the repetitions do give the impression that the parrot is interested in duetting with the human singer. Even some of its bodily attitudes appear to me strikingly attentive, and they do display a dynamical physiognomic quality in its movements. Something very interesting is related to the pattern of the phrase 'Ti voglio'. While Eric spells the syllables distinctly, Pavarotti moves quickly from the initial 'Ti vo', gliding fast upwards to the 'glio' where the pitch settles. In fact his diction is less than perfect in this passage. He slides almost continuously instead of demarcating the first two pitches. This is not a criticism, of course it is a matter of choice and style that the great tenor has made. But it must be emphasized that here Eric is very creative. Throughout the performance Pavarotti does not reproduce the passage in question much differently than he does at the beginning (though the song is rather slow and repetitive). Eric offers an interpretation of the song, not a parroting of the song. In this regard we should revise and reconsider the meaning of the verb 'parrot' accordingly. This question of interpretation vs. repetition is striking. To reiterate, I have the feeling that the bird is 'in the performance', it believes in its own contribution and creative participation. Eric is strongly interested in evolving the performance together with the leading tenor. It would be fascinating to know if there was an evolution over time in the style of Eric's performances.

Resonance is a technical concept that I have explained in a very simplified manner above. In a mechanical context it presupposes that bodies that become resonant have some intrinsic vibratory characteristics in common that allow them to be tuned to one another when one of them starts oscillating at the shared frequency. But the idea can be explored further. I would make the suggestion that states of resonance may appear under similar circumstances in diverse phenomena such as biological metabolism, behavior, performance, and cognition. Eric and Pavarotti do seem to have entered into a state of resonance. Eric has discovered that it possesses common frequencies that it exploits in order to contribute to the performance creatively. This fantastic performance is an attempt at a complex communication with the other who is singing along. Unfortunately it is unidirectional in that Eric interacts with a video recording and not with another singer in a live situation. We can speculate about the wondrous possibility of a live situation in which Eric might become part of the flow of events. In this hypothetical performance we might imagine an improvisation between Eric and Pavarotti. Improvisations are communicative and creative in subtle ways that have been studied by musicologists. Drawing on a comparative study of musical performances in different cultures, Bruno Nettl advanced the hypothesis that improvisation and composition are not opposed concepts, and suggested that we should not juxtapose them, but rather see improvisation and composition as part of the same idea.[61] Keith Sawyer proposed that instead of considering composition/creativity and improvisation as being the ends of a two-dimensional space, different types of performance should be contrasted along different dimensions, so that composition/creativity and improvisation represent "points maximally opposed in a multidimensional space".[62] According to Sawyer, the maximally opposed points of the multidimensional continuum are more appropriately considered to be 'ritualized performance' and 'improvisational performance'. This means that improvisation is not an on/off quality and that we should rather think in terms of a degree of improvisationality, or extemporization, determined by a point in the multidimensional space defined by multiple contrast dimensions (which here I cannot discuss in more detail). The notion of multidimensional space leads us to introduce the concept of emergency in improvisational settings. Thus, for Sawyer, improvisational creativity can be characterized as follows.

> Performers are constrained to operate within the performance genre. A given act is more narrowly constrained by the emergent. The nature of this constraint is unique and specific to the performance and the moment of the interaction. In improvisational genres, each performer is expected to contribute something original to the evolving emergent in each act, through the process of indexical entailment. In the choice of indexical entailment, performers are subject to the constraints of the emergent. In response to the performer's action, the other participants evaluate the act, and the subsequent interaction determines to what extent the indexical entailment resulting from the act affects the (still/always evolving) emergent. This 'evaluation' is often immediate and often not consciously goal-directed. A more skilful indexical entailment is more likely to enter the emergent, thus operating with more force on subsequent performance acts.[63]

These elements of the emergent situation in an improvisational setting can to some degree be ascertained in the duet between Eric and Pavarotti. Eric is attentive to the cues that the other singer is giving, he clearly learned the patterns of the song in previous rehearsals and deploys its knowledge in creative ways that it thinks to be appropriate to the flow of events. The evaluation of the indications given by cues is not consciously made or goal-directed. Typically goal-direction is not a strong constraint on improvisational performances, as opposed to performances based on written compositions in which directions and guidance are generally given to the performer by the composer. The former is obligated to respect the performative directions and has little freedom to introduce personal variations. In improvisational situations communication operates under different circumstances. It is more like speech in everyday dialogic settings in which no written or premeditated text is guiding the discussion. The interlocutors approach one another on the basis of certain expectations, or chance events such as bumping into each other, and are open to certain cues that derive from other participants and which may of course be vastly different. They might include verbal behavior, memory, emotions, or body language, gesture and other bodily and environmental cues. Communication flows in any case. While singing the performers communicate under a holistic state of resonance. This state is what Jazz players often indicate with the weasel word 'swing'. Being into the 'swing' of the performance, what is it? It seems to me that holistic resonance comes very close to what is meant by them with this mysterious concept. There is swing, or there isn't, as they say, that's all. If there is no swing, the performance fails in that the players have not really entered into it.

The take in the recording studio may be discarded by the producer. But if there is swing, then the players might say something like the following, 'I was really into it', 'I felt its qualities', 'I was one with it', or such like phrases. Interspecies artistic communication might happen on the basis of holistic resonance under certain circumstances. *Papparotti* is real, artistic communication might happen. Perhaps this would be as flabbergasting as encountering the extraterrestrial aliens, they might be with us already, we just didn't notice.

4. The Epiphonetic Landscape

Although musical notation gave Luigi Paolucci an innovative and versatile toolkit for rethinking the nature of birdsong, he acknowledged that it could not capture the variety and complexity of the timbre structures that are commonly heard in birdsong. He introduced a conventional symbolism that simplified the transcription by apposing letters to the musical staff in order to indicate the four fundamental timbres that he had recognized in birdsong. They are the *supple*, the *trilling*, the *harsh*, and the *acute*. What I find extraordinary is that Paolucci developed a concept of music in birds that is based essentially on the extension and structure of the harmonic content. The timbre of the bird's voice is the fundamental quality of birdsong. It was not so much pitch that defined the musical qualities of birdsong but rather a more complex structural component of sound perception. The significance of timbre in psychoacoustics and music perception theory became very clear around middle of nineteenth century especially in the wake of the work of Hermann von Helmholtz, on which Paolucci relied.[64]

The study of the mechanics and neurophysiology of the sense of hearing became a research field in its own right precisely around the middle of the nineteenth century. The foundations of the field were laid by physicists such as Georg Ohm and Hermann von Helmholtz and by the physiologist Alfonso Corti. Ohm and Helmholtz realized that importing Fourier analysis (a mathematical technique for representing periodic functions as sums of sinusoids) into the field of hearing research would pave the way for a new understanding of how we perceive the world of sounds. But it was Helmholtz that coupled that insight with the new discoveries by Corti, thus formulating a theory of hearing whose general thrust dominated the field for many decades. Indeed, there is little doubt that Helmholtz succeeded in combining two strands of research that until then had remained separate, namely mathematical acoustics and the physiology of hearing. But the impetus to integrate the two came from the anatomical–physiological researches of Corti. In 1851, Corti's seminal paper announced the discovery of a structure in the mammalian inner ear that later on would be called organ of Corti in his honor. In the subsequent decade Helmholtz published the first edition of his celebrated book *On the sensations of tone*, in which the theory of the ear as a frequency analyzer was presented. Corti's paper was not only a milestone in anatomical–physiological research; it suggested the crucial analogy that Helmholtz would masterfully develop into a full-blown theory of human hearing. To illustrate his findings, Corti drew three floating hammers above the basilar membrane (a complex structure in the inner ear of the mammal cochlea) that he thought could strike the membrane in response to an external stimulus and thus excite the nervous terminations, which are the expansion of the acoustic nerve in the cochlea.[65] Thus, the hammer mechanism of the piano keyboard, the most popular instrument of the Romantic epoch, suggested to Corti the perfect analogy for the mechanics of hearing. Helmholtz refined Corti's explanation by combining it with Ohm's theory that sensations of pure tones are evoked by sinusoidal pressure waves.[66] In effect, Helmholtz postulated that the mechanism for frequency analysis in the organ of Corti must be sympathetic resonance. In other words, the ear performs a Fourier analysis of the incoming external stimulus and resolves its constituent pure tones. However, as he conceded in the fourth and final revision of his book, the anatomical mechanisms of the ear that sympathetically resonate in response to the pure tones compounding the external stimulus could not, for the time being, be identified with certainty.[67] By the time he published the fourth edition of his book, in 1877, more detailed knowledge of the organ of Corti had replaced Corti's romantic floating hammers with less inspiring arches, which were not free to float but attached to the surrounding anatomical structures. Corti's brilliant analogy turned out to be untenable but the thrust of his argument has survived until the present time. International conferences on the mechanics of hearing and psychoacoustics show abundantly clearly how the Corti-Helmholtz theory of frequency analysis in the ear informs current research on the biomechanics of hearing. Indeed Helmholtz's perplexity about the parts of the ear responsible for resonance has proved prophetic. Corti's analogy with the piano hammer mechanism implicitly solved the problem of resonance by alluding to the *pectinate* zone as a series of densely packed piano strings. Each string would naturally resound to the proper frequency to which it was tuned, and being located at a particular place along the basilar membrane would naturally establish a topographic organization of resonance. This topographic organization is commonly referred to in the literature as *tonotopic* organization. More precisely, it refers to the spatially distributed frequency–sensitivity of the sense of hearing. Unfortunately, with the displacement of Corti's romantic analogy and the appearance under the microscope of the more mundane arches, it became difficult to locate the tuned resonators inside the ear that could perform the task of Fourier decomposition of the incoming acoustic stimulus. Nowadays the phenomenon of resonance and frequency analysis in both human and non-human hearing neurophysiology is conceived of in a much more sophisticated manner than Helmholtz could envisage. But it was Helmholtz in the wake of Corti that inaugurated a revolutionary conceptual framework in which perceptual and cognitive structures of sound and mind might be theorized and investigated empirically. Paolucci fully realized the momentous import of Helmholtz's researches and intuited the richness and variability of timbre patterns in birdsong.

In the Corti-Helmholtz conceptual framework the perceptual attribute that humans generally recognize in sounds and label 'timbre' is explained by the structure of the so-called harmonic content of the sound.[68] We have already seen an example of this phenomenon in the spectrogram of the duet between Eric and Pavarotti. In these diagrammatic representations the harmonic content appears as series of lines vertically stacked, with different levels of intensity and somewhat regularly separated from one another, and also evolving horizontally along the temporal axis. It is the distribution of energy along the vertical axis of frequency that determines the aural qualities that we summarize with the label 'timbre'. This is certainly a simplification. In reality complex and subtle relationships among psychoacoustic percepts and their temporal dynamics interact continuously in forming the sonorous image of the sound. The timbres of Eric and Pavarotti are different because they have different harmonic contents. Their timbres give rise to different concentrations or intensities of energy (typically represented with color or denser grey tones in these diagrams) in the series of lines vertically arranged that may or may not overlap. But it is extremely important to realize that these diagrams or spectrograms do not give us an *objective* picture of the sound structure. It is a misconception all too frequently popping up in the popular literature (and very often even in technical scientific articles) that frequency analysis of sound is an objective representation of the character of the sound. The misconception originated in the accidental fact that until the end of the nineteenth century the only set of functions that could be arranged to represent more complex periodic functions such as sound waves (which could be seen as summations of the simpler components under opportune circumstances) was the Fourier set of sinusoids. In the early twentieth century new systems of so-called orthogonal functions were discovered that could do the job even better then the Fourier set. We now know that there are infinite sets of such functions that can represent complex periodic sound waves. We are spoiled for choice. In principle each system would produce its own spectrogram that would be strikingly different than the Fourier spectrogram. Hence there is no such thing as an objective representation of sound to be extracted from these mathematical techniques. They are simply means of visualizing the sound wave in different ways that may be more or less convenient for certain purposes such as telecommunication or digital technologies. We have become so accustomed to the language of pitch and timbre that we tend to forget that these constructs do not have an objective basis in the spectrograms. They have a historical, intellectual and social importance because these perceptual attributes of sound and music were instrumental in forming the modern tradition of Western classical music that has dominated the way in which music has been produced and consumed in Western culture for the last few centuries.

Paolucci's indication that the fundamental character of birdsong is not so much pitch but rather timbre must be seen in the context of the above discussion. His insight shifts the focus of attention from pitch to timbre, and gives us the opportunity to integrate our understanding of birdsong into a larger cognitive and philosophical perspective. The fact that sound waves can be 'seen' from strikingly different angles thanks to multiple systems of sound representation opens up a new vista. It suggests that sound perception can operate radically differently in different animal species such as humans and birds, and that birds may form a sonorous image of the world that might be substantially different than ours. That is, it might work more like another type of spectrogram. I will mention one beautiful example to clarify matters a little more. At the beginning of the twentieth century the Hungarian mathematician Alfred Haar discovered a set of functions that have since been named 'Haar functions' in his honor, and which would produce a spectrogram that has totally different characteristics from the Fourier spectrogram.[69] I will only offer a qualitative description and avoid the technicalities. Instead of sinusoid functions that are characterized by amplitude and frequency, the Haar system includes functions that look more like impulsive bursts. They can be grouped together following the numeral order of the integers 1, 2, 3, 4, etc. To each number there corresponds a set of 2^n functions. (So for n=2 the system includes four impulses, for n=3 the system include eight impulses, and so on). All impulse functions have the same periodicity as the target sound wave. So if the latter is a sound pitched at A (440Hz), then all impulses would be pitched at the same exact value. This is the most striking feature of this system. It means that although the component impulses are periodic, they are all set at the same pitch as the target sound, unlike the harmonics which instead progress following the number 1, 2, 3, etc. So in the Haar spectrograms there are no harmonics at all! The Haar spectrogram is a representation of the sound without harmonics. All pitches are the same. There is no obvious way of describing the character of the sound represented by a Haar spectrogram. It does not have a harmonic content and yet there is a timbre and a pitch periodicity if the sound elicit these perceptual attributes. This fact demonstrates quite well that it is totally arbitrary to associate the perceptive attribute that we call 'timbre' to the harmonic content of the sound. Timbre is not the perceptual correlate of an objective structure in the sound called 'harmonic content'. The latter is simply one way of visualizing what we perceive as timbre. But it is crucial to remember that it is just one of the infinite possible ways of doing this visualization. In addition to all this, one has also to realize that once the arbitrary and conventional nature of the sound visualization through spectrograms has been understood, there is no reason to presuppose that the same sound phenomenon must give rise to the same perceptual attributes in humans and non-human animals. There is probably nothing really new in this conclusion, but its motivation in the analysis of sound through spectrograms clarifies the real problems. I am not suggesting that we should expect birds to perform in their brains complicated analytical operations that allow them to operate like Haar spectrograms. The analogy that the brain and the mind operate like a spectrogram was the fundamental idea that Helmholtz inaugurated in the wake of Corti's discoveries. It is a brilliant idea rooted in an analogy. Hence there is no reason to believe that animals perform spectrograms. What I have suggested is that by understanding the possibility of multiple spectrograms, we come to understand that the perceptual attributes of sound (and in the final analysis the consciousness of sound) are phenomena that can be illuminated by analogies, even though we might never be able to enter into the mind of a bird to hear the world in the way that birds hear. The Haar spectrogram makes it clear that the acoustic world might be perceived in terms of its timbre structures as a superimposition of different timbres. While the Fourier spectrogram emphasizes the attribute of pitch, the Haar spectrogram emphasizes the attribute of timbre. Both are granular, atomistic, and discrete representations. In the former, a sound is conceived of as being the summation of many sinusoids (often called 'pure tones') whose pitches progress in the order of the natural numbers, and whose timbres are the same. In the latter, a sound is conceived of as being the summation of many impulses whose pitches are the same as the sound but whose timbres are all different according as the impulse belongs in any of the groups progressing as the number two raised to the natural numbers. We might say that they give us two correlative visions or models of the acoustic scene. One is pitch oriented while the other is timbre oriented. On the basis of Paolucci's research and his insights into the four fundamental timbres of birdsong, we might conclude that the acoustic world-view of humans may be more pitch oriented while that of our fellow birds is more timbre oriented. I find this insight by Paolucci into the life-world of birdsong to be perhaps the most precious, philosophically pregnant and fascinating to be found in his essay.

Recent research into the mechanism of hearing in birds offers more evidence that pitch might not be as salient in birds as it is in humans. In birds, "[o]ne of the most basic descriptions of auditory capability is the audiogram, that characterizes the frequency dependent variation of absolute sensitivity. Partly due to the columella type single ossicle middle ear of birds, sound transduction deteriorates above 4 kHz and the high frequency hearing range in birds is limited to frequencies at or slightly above 10 kHz".[70] These two facts, i.e., that the frequency range is limited and frequency discrimination deteriorates above 4Kz, should be considered in the light of the convergence characteristics of the Fourier spectrograms. Basically, in order to reconstruct a periodic function by means of summations of sinusoids pitched at progressively higher and higher frequencies, a large number of sinusoids is needed. Practically this implies that if the hearing apparatus is unable to perform frequency discrimination above a certain limit with sufficient precision the mechanism of the Fourier spectrogram is precluded and/or unable to work properly. As birds cannot discriminate well above that range, they cannot perform Fourier spectrograms within their neurophysiologic apparatus. This limitation raises important questions concerning the tenability of the Helmholtz-Corti conceptual framework. It is doubtful whether the Fourier spectrogram is the correct representation of sounds from the perspective of birds. The identity of a sound that cannot be represented with all the necessary Fourier components fades, that is, the sound becomes blurred and can be confused with others. It loses its signature. And the Fourier signature of a sound is very complex as it takes many sinusoids to converge to the sound. If birds can hear sounds with sufficient precision for them to recognize their species calls and songs and possibly other sounds, they must rely on mechanisms that maintain the signature of a sound intact. Moreover, the complexity of the spectrograms of birdsongs (we shall presently see the example of the nightingale) indicates that birds must be able to decode the timbre quality of sounds very precisely. Their voices are very complex. They are not simple sinusoid-like sounds. It is true that certain aspects of both the anatomical structures and the physiology of the hearing apparatus of birds resemble those of the organ of Corti in mammals. But there are differences that are not presently understood.[71] It is conceivable that birds should have developed specialization mechanisms for discriminating and appreciating the sounds that they produce. Thus, their hearing apparatus must be sufficiently complex and efficient in order to respond to the complex sounds that they produce. Birdsong patterns and bird vocalizations are rich in dynamic variation. Again, the slow convergence rate of the Fourier spectrogram is unlikely to be able to represent this highly dynamic complexity. In sum, we must be cautious in extending to birds the idea that they cognize sounds as if they were able to perform the Fourier spectrograms of the sounds. On the other hand, Haar spectrograms have more rapid convergence characteristics, which make them more suitable and reliable (in the presence of noisy environments, for instance) for efficiently representing complex sounds in terms of impulse-like functions. This means that they might work better from the perspective of birds because these spectrograms require processing fewer components in order to reconstruct the signature and therefore maintain the identity of a sound. A Haar spectrogram can be thought of as a representation of the sound which has on the horizontal axis the dimension of time along which the sound evolves and on the vertical axis a succession of timbres, the resolution of which grows as the number two raised to the integers in the series of the natural numbers, that is, 1, 2, 4, 8, 16, 32, 64, 128, etc. Each timbre is progressively more and more defined in term of resolution as it is determined by a correspondingly larger number of parameters. So, for example, the second timbre has four parameters, the third eight, the fourth sixteen, the fifth thirty-two, and so on. The third dimension indicating the values of the parameters can be represented by color or grey intensity as in the Fourier spectrogram. This insight into sound representation suggests that sound can be explored as having an evolving dimension of timbre complexity rather than an evolving pitch like in melodic sequences. More precisely, we might eventually rid ourselves of the notion of timbre altogether, a category which after all remains squarely dependent on human theoretical, acoustic and esthetic preoccupations. We do not have a proper terminology for the evolving morphological structures that appear in Haar spectrograms as they transcend the correlate notions of pitch and timbre. Unfortunately, Haar spectrograms are not commonly in use in bio-acoustics. (The traditional Fourier spectrograms are much more easily available.) But what matters most is that they help us shift the focus of the imagination from the traditional categories of human music towards more alien acoustic scenarios. I call these scenarios 'epiphonetic landscapes' (as I will explain in a moment). Hence in the remaining portion of this section I will return to the Fourier spectrograms of birdsong but from now onwards I will focus on their features from the viewpoint of the epiphonetic landscape. Hopefully, this radical shift in our focus that has been motivated by the Haar spectrograms will allow us to catch a glimpse of the sonorous images of the acoustic environment that birds themselves might form in their mind.

First of all, with the help of an exquisite example I will proceed to illustrate the four fundamental timbres that Paolucci perceived in birdsong. I will use a segment from a nightingale song recording.[72] Obviously, as Paolucci did not have at his disposal modern technology, he could not perform spectrograms of bird-songs. Although his analyses may have been limited, in that they were based on observations with the 'naked' ear, his insight that timbre is a defining feature of birdsong suggests that we should look for fine grain timbre-like patterns in the spectrograms and be attentive to the morphology of the sound phenomenon as a whole without succumbing to the influence that Western science and music concepts have inculcated in our thinking habits. With the advent of the spectrogram in the twentieth century we have entered the era of 'telescopic' psychoacoustics but we need to dissolve disciplinary boundaries and couple the power of technology with the philosophical spirit. Perhaps this strange adventure in avian hearing will reward our attempts to understand and respond to the ethical questions that demarcate humans from non-human animals. It will thus allow us to inaugurate a genuine human/animal psychoacoustics more than century after the pioneering work of Helmholtz and Paolucci.

The Epiphonetic Landscape Explained

With this expression I designate the ensemble of phenomenological structures that emerge in birdsong spectrograms but which cannot easily be categorized within the conceptual frameworks of linguistics, music, and human psychoacoustics. I suggest *epiphonetic* in analogy with *epigenetic* where the transmission of traits and information to the next generation of living organisms is considered in the light of biological mechanisms and cultural possibilities that question and transcend the materialistic dogma of genetics. In their book on the history of life, Eva Jablonca and Marion Lamb convincingly argue that "there is more to heredity than genes; some hereditary variations are nonrandom in origin; some acquired information is inherited; evolutionary change can result from instruction as well as selection".[73] In the same vein, I would like to point to hermeneutic possibilities that transcend the dogmas of phoneticism, which are rooted in the nineteenth-century positivist conception of science. Music has often been regarded as a language but this worn-out metaphor, no matter how illuminating it might have been in its youth, has its own drawbacks. It introduces phonetic and linguistic categories in discourses about animal music and vocalizations that in the end might impede our living-through the experience of animal sonorous life-worlds, including non-human esthetics and life and inter-species communication.

The human perception of the acoustic scene is temporally dilated in comparison with that birds. This surprising phenomenon must be consciously factored in when we make an effort to imagine in our minds the sonorous world of birds. This is clear from the spectrograms. We observe that in performing the calculations necessary to construct the spectrogram from sound data, the scale of the temporal window that shifts along the temporal axis as the sound evolves in time must be decreased significantly in order for the acoustic structures to come into focus. This important phenomenon tells us that the acoustic worldview of a species that relies on these structures emerges at a different temporal resolution. In other words, when we humans listen to sound or music, we construct a moving image of the acoustic phenomenon whose salient features (such as the evolving dynamics of timbre and pitch) manifest themselves at a temporal resolution that depends in part on the human hearing system. We proceed in our hearing more slowly than birds. If we were able to replace our hearing neurophysiology with that of a nightingale, we would perceive the acoustic scene as if it were blurred in both pitch and timbre. This blurring effect is not easy to imagine. Acoustic blurring is a smearing of the pitch and timbre of a sound that makes it less definable. It is not a glissando or a metamorphosis of timbres but a sound that tends towards what we perceive as noise. Regardless of species specialization, however, there is also a fundamental limit in the trade-off between the desirable features of both high resolution of the acoustic scene (that is to say focus) along the temporal dimension and high resolution along the pitch scale. The more focused the acoustic scene becomes in terms of pitch, the more diffused or blurred in terms of time. All species must submit to this constraint that does not depend on the specialization of their biological mechanisms but on mathematical-physical laws that appear to hold true for acoustic phenomena. Different species may have adapted and explored sonorous life-worlds focused on the different points of equilibrium at which they have found it convenient to position themselves. The finer the resolution in time, the more blurred the resolution in pitch, and all the infinite gradations in between are as good as any. For example, a human melody cannot run faster than the limit that allows the pitches to remain sufficiently distinct for human ears. All sounds tend to be perceived by humans like noisy clicks as they become faster. It is a consequence of the same constraint on the trade-off between pitch and time. Fast sounds are perceived as clicks because we cannot resolve their acoustic complexity into pitch and timbre. Another species might have no problems. Birds would do a better job at fast melodies if they were interested in melodies. It is highly unlikely that they are, though. As I have suggested, in our parlance we should say that they are more timbre-oriented creatures. In Figure I.4 we see a spectrogram of a short section (about four seconds) from the nightingale song mentioned above in which the trilling and harsh timbres identified by Paolucci are very prominent.

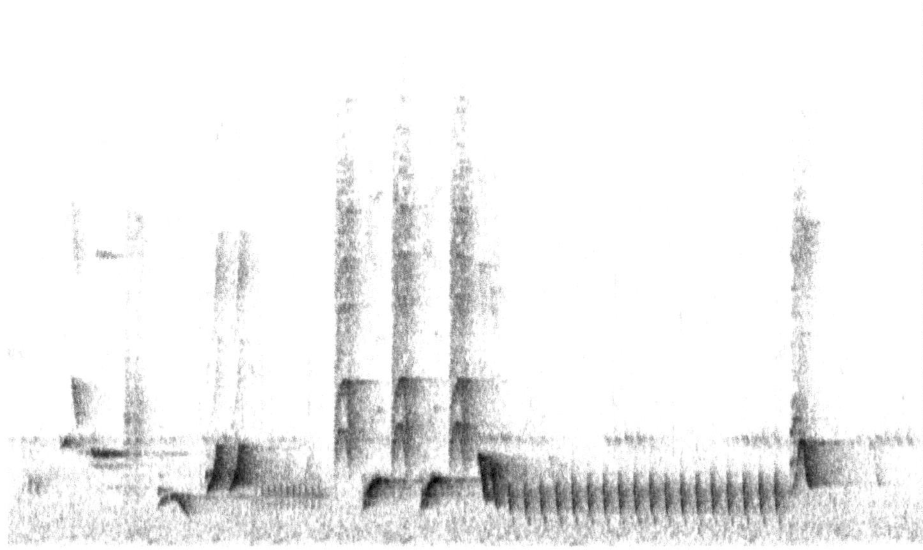

FIGURE I.4 The spectrogram of a fragment of a nightingale song. Time evolves from left to right. Frequency increases from bottom to top. The vertical structures jutting out upwardly correspond to what Paolucci called 'harsh timbre', while the densely packed area of repeating short lines on the right hand side (close to the bottom of the picture) is the trilling timbre. The other two timbres that Paolucci described are also heard in the song though they are not shown in this section of the spectrogram. They are morphologically simpler and they tend to sound like flutes or piccolos playing at low and high pitches.

I will now develop by thought-experiment an imaginative reconstruction of the acoustic scene represented in the spectrogram. I will do so from the *perspective of the nightingale*. At this point, I must leave behind all traditional music categories such as pitch and timbre for which there is no a priori correspondence in the acoustic life-world of the bird. The spectrogram will serve only as a rough guide into an uncharted territory. The future possibility of more accurate spectrograms based on multiple functional systems such as the Haar impulses will allow us to make the necessary corrections and to add precision, intelligibility and finer details to the thought-experiment. We begin by considering the spectrogram as an acoustic three-dimensional space, in which the intensity of the grey tones should be mentally pictured as the heights of regions located at certain positions in the plane of the figure. The objects in this acoustic space will then have a three dimensional morphology. Darker regions will correspond to higher peaks and lighter regions to deeper valleys. The prominent vertical structures in the figure have an interesting characteristic in common. Their left border is more defined, indicating a marked attack, while they tend to become blurred as they fade in time. In the three pillars at the center we see three arch-like shapes showing two narrow arches above one another (somewhat like in a Roman aqueduct). They also look like narrow but well rounded capital *A*s. The contours of these marked regions are generally interpreted as pitch-like sounds, or quick glissandos, for in effect we hear them sounding somewhat like this even though they are accompanied by hissing noises and more consonant-like sounds. But this is the human acoustic viewpoint that has settled at a particular equilibrium in the trade-off between time and pitch. As the nightingale listens to its song patterns sculpting the sonorous matter, it would discriminate the blurred areas much more creatively. It would form an impulse-like acoustic picture, very rich in sculptural textures, which is not noisy but densely packed with microsounds that we cannot possibly imagine (since unfortunately our hearing blurs the acoustic scene at this scale). In the nightingale's acoustic life-world birdsong materializes as a temporally evolving sculpture, which is made of a magical element that is conjured into existence by singing. Singing is like a magic wand that the bird waves to orchestrate the sonorous world into existence. While singing up the world, the sonorous sculpture takes shape like a living architecture that is continuously modified, let fade into silence, and evoked again in myriad variations. These sonorous architectures shape the life-world in which song birds such as the nightingale inhabit. As the equilibrium in the trade-off between time and pitch scales upwards towards a region of lower temporal definition, we can speculate that avian hearing would metamorphose into more human-like hearing, and that in consequence the three-dimensional sonorous sculptures would gradually lose their higher definition. They would begin to appear more like out-of-focus lumps and/or confused aggregates, as we see in the spectrogram, heavy and more inert, taking on the slower-tempo qualities of the familiar perceptual attributes which humans call 'pitch' and 'timbre'. Whereas humans are grounded, birds are creatures of lightness, beings that have annulled the force of gravity. Hence flying must be a crucial determinant of the worldview of the bird. Its physiology is in effect part of the same physiological system of breathing and singing. The nightingale sings, breathes, and flies its world into existence. In the human experience, sound emerges from the silent past and continually fades into the silent past because we are slow, earth-bound creatures, dragged down by weight towards the silent bowels of the earth. Human music-making and sound project forwards, like missiles thrown into the future, as if we were attempting to escape gravity, playing a ballistic game with an artillery of music instruments. Yet inevitably sound bullets fall back into the earthy, silent past. To a creature like the nightingale that has escaped the bond of gravity time must appear like the esthetic fourth-dimension of life. A creative dimension naturally built into its own physiology, in which singing breathes the architectures of song in and out of existence at a fast tempo.

By exploring the epiphonetic landscape of birdsong we have caught a glimpse of the life-world of the nightingale. We have opened up a new vista, discovering that a continuum of gradations, possibly not yet punctuated by any animal species (or once inhabited by extinct ones), separates the acoustic equilibrium at the scale of the singing birds from the musical humans.

In the light of the epiphonetic landscape, both human music and language, that is, phoneticism, appear as a possibility inscribed in biological life and emerging at one precise point along the continuum of scale equilibriums that abide by the natural law of the trade-off between temporal and pitch resolution. We have sung and spoken (at least in the sense of producing articulate mouth-noises) since we came to punctuate this bioacoustic continuum at one particular position. We cannot characterize birdsong as 'song' in this strict human sense, insofar as bird singing does not emerge at the same position along the continuum. The psychoacoustic 'other' resists assimilation into phoneticism, stands up to the imperialism of the articulate voice, and we do not have a proper vocabulary for the phenomenon of birdsong. We are phonetic creatures but phoneticism as such does not define the life-world of the singing birds. As we move along the continuum of scale equilibriums, we encounter infinite acoustic life-worlds that may not have been experienced in the course of biological life. These possible acoustic life-worlds are like virtual beacons in the future of biological existence, or perhaps deserted signposts from the past.

At this point we can deepen the thought-experiment in bird hearing quite considerably. We must call into question the theoretical foundations of the spectrogram. Amazing as it might seem at first glace, we shall see that there is one fundamental assumption behind the spectrogram that is untenable. The mathematical techniques that allow us to construct the spectrogram and the neurophysiologic interpretation of the spectrogram (in essence the Helmholtz-Corti framework) are two entirely different concepts. There is little or no evidence that hearing in humans and birds works like a spectrogram although this dogma has been believed since Helmholtz. The constraint of the equilibrium between temporal scale and pitch that we have discussed so far is simply a property of the mathematical expansion of the sound wave in terms of systems of orthogonal functions such as the sinusoids. The acoustic interpretation of this constraint suggests the physical law that I have so far assumed as valid, in harmony with the Helmholtz-Corti framework. This physical law, however, need not be reflected in the biology of the living organism. There is no a priori reason why a living organism that has developed hearing should abide by a physical law which is rooted in the phoneticism that seems to be a prerogative of humans and that, as we have seen, has turned out to be very much doubtful in birds. That the Helmholtz-Corti framework is inadequate to investigate the complexity of sound cognition in living organisms can be suggested in different ways. We will discuss salient cases. Let us consider the issue of the convergence of a Fourier representation of a sound. Fourier theory states that any periodic signal such as a stationary sound wave can be represented by an infinite series of sine waves, commonly referred to as pure tones. When we perceive different sounds, we perceive their different characteristics because our hearing apparatus, namely, the external, middle and inner ear, altogether is believed to perform a Fourier analysis of the incoming sound and makes available this information to the brain. However, since we cannot perform infinite operations, the Fourier analysis performed by our hearing apparatus can only approximate the sound with a finite number of sinusoids. But only with an infinite number of sinusoids does Fourier theory guarantee the so-called point-wise convergence of the component sinusoids to the periodic signal. Thus, when we hear, we never point-wise converge to the external stimulus in a mathematical sense because we can only add so many sinusoids. We only converge approximately. This fact is of fundamental importance. It allows us to make an empirical prediction. First of all, consider this procedure. Extract from a sound its Fourier components, or synthesize the sound by so-called additive synthesis (by adding a number of component sinusoids together). Subsequently, approximate the sound by adding only so many components together—say, one, ten, fifty, let's say n components, and so on, creating many different sounds according to the number of components (call each of them *soundn*). This way we can produce many approximations of the original sound, listen to them, and find out how many components are necessary for us to hear exactly the same sound as the original sound. Now, the prediction is the following. If our hearing apparatus works as a Fourier analyzer, a sound approximation, *soundN*, will always be possible which is compatible with human neurophysiology and sounds like the original, i.e., we do not perceive any difference between *soundN* and the original sound. Let us call the perception of *soundN* "perceptual convergence" to the original sound. In other words, any sound and its perceptual convergence will sound the same to human listeners. If human listeners are Fourier analyzers, they must resolve the necessary and sufficient components to achieve perceptual convergence. How many Fourier components can we meaningfully add together in order to generate *soundN*, the perceptual convergence of a sound? Let's see. There are two constraints we must take into consideration, one mathematical, one neurophysiologic. The mathematical constraint is as follows. Fourier theory states that all of the components of a periodic sound will have frequencies that must be multiples of the so-called fundamental frequency; for example, if the original sound is that of an oboe playing an A at 440Hz (the classical orchestra tuning note is generally sounded by an oboe), all components of the sound produced by the oboe will have frequencies that are multiples of 440Hz (for example, 880Hz, 1320Hz, etc.). Since it is generally assumed that the upper frequency limit of human hearing is 20000Hz, we can only hear components whose frequencies do not exceed that limit. Hence we can meaningfully add components to our approximation that are below that upper limit, otherwise they will have no perceptible effect on us. For example, if the fundamental frequency of a sound is 400Hz, the maximum number of meaningful components we can add together to achieve perceptual convergence, *soundN*, is 50 (since 50x400=20000); thus N=50. The physiological constraint is as follows. There are approximately 3500 receptor units within the organ of Corti in humans. If a receptor unit, or a cluster of units, codes information relative to the frequency to which it is tuned (as is customarily assumed), there will be up to 3500 units available along the basilar membrane for coding frequencies (though their frequencies do not uniformly span the audible spectrum). If a cluster of unit codes for the same frequency, then fewer coding units will be available. So, the perceptual convergence of a sound could at most have 3500 components, which gives N=3500 as a maximum value for N. In the above example, the figure of 3,500 receptor units largely exceeds the number of meaningful components (50). But think of a sound such as a square wave with a very low fundamental, say 1Hz, then up to 20,000 mathematically meaningful components are possible, possibly requiring 20,000 receptor units, a figure which is way above the limit of the physiologically maximum value N=3,500. In sum, the Fourier analyzer producing the spectrogram within the human hearing apparatus, if there is any such thing, must meet the requirements of both the mathematical and physiological constraints, which limit the number of meaningful components that can be added together in order to achieve the perceptual convergence of a sound. Easy psychoacoustic experiments show that certain sounds can indeed perceptually converge, i.e., the perceptual convergences of such sounds do sound to human listeners like the originals. For example, a square wave whose fundamental frequency is higher than approximately 5Hz can indeed perceptually converge. However, other sounds such as that of a square wave of very low fundamental (1Hz, say), or the sound of the word 'brilliant' produced by a human speaker, cannot perceptually converge. In other words, no matter how many meaningful Fourier components we add to it, the approximation never sounds like the original. Yet we hear the original sound! How can we do that? We must conclude that our hearing apparatus can do much better than produce a spectrogram by Fourier analysis. In fact, we perceive sounds which cannot be approximated to a sufficient degree of perceptual accuracy by a Fourier analyzer compatible with the neurophysiology of human hearing. Briefly, we hear sounds that do not perceptually converge. The reasoning holds true for birds by analogy as their system includes a system of receptors similar to that of humans. Hearing does not work like Fourier analysis. Biology knows much better.

Another example is this. Let us consider the time-direction of sound perception. Can you imagine how the word 'brilliant' will sound like if pronounced backwards? Hard to imagine, since by backwards I do not mean to try and pronounce it phoneme by phoneme *à rebours*. I really mean backwards in time, instant by instant. But it is not hard for a computer to produce the really inverted sound of 'brilliant'. Indeed, the sound of the backward 'brilliant' turns out to sound really, really different than that of 'brilliant' (very much unlike that of 'tnaillirb'). So much so that you could hardly guess the original word when listening to its inverted waveform for the first time. But Fourier theory implies that the spectrum of any periodic waveform, its atomic signature, is identical regardless of the direction in which time flows; that is, the spectrum of 'brilliant' is identical to that of its temporally inverted waveform. In other words, if our hearing apparatus worked as a Fourier analyzer, we would be unable to distinguish between the sound of 'brilliant' and the sound of its temporally inverted waveform. Yet they sound to us so different (the reader can try this simple exercise on a laptop computer). How can that be? We are now in a position to doubt the initial claim that all species must submit to the equilibrium constraint depending on mathematical-physical laws that appear to hold true for acoustic phenomena. These mathematical-physical laws are the result of a long historical process of development that has given rise in Western Europe to a materialistic system of beliefs that we nowadays call 'science'. Yet science is deeply rooted in the dogma of phoneticism. There cannot be science without phoneticism and its attendant dichotomies such as vowel and consonant, and it cannot be an accident that modern science emerged in one civilization only, one whose fundamental languages such as Greek and Latin are rooted in the phonetic dualism of consonant and vowel. But the idea of the epiphonetic landscape transcends the materialistic dogma and the phonetic dualism, and points us in the right direction. Hearing in living organism need not be thought of as phonetically dichotomous. Hence the apparent limitation imposed by the law of scale equilibrium between time and pitch resolution turns out to be an illusion that the scientific belief system has created. Species might indeed hear at different temporal scales without violating any such illusions as illusions are not unbreakable barriers. The spectrogram itself is the reification of an illusion. The epiphonetic landscape is real but need not be made *visible* in living organisms by cognitive spectrograms of the sound waves. Rather it is made *hearable* by the living organism through hearing processes that science itself will be unable to clarify in principle, as long as it remains anchored in materialistic assumptions that are deeply rooted in the dogma of phoneticism. Ever since Ohm introduced it, hearing scientists have accepted Ohm's acoustic law. It was Helmholtz's genius that allegedly solved the mystery of how the Fourier spectrogram obtained by analysis is eventually reassembled—or rather forgotten about—by the human mind in order for humans to perceive sounds and not jumbles of sinusoids. Helmholtz's solution was his theory of signs. The human mind makes use of items of acoustic information as signs of external-world objects, that is, only to the extent that such signs serve the practical purpose of tracking down objects belonging to the external world. In other words, sounds are acoustic signs of special objects existing in the environment. Fourier spectrograms are irrelevant for the purposes of cognizing the environment, and hence, according to Helmholtz, mind forgets about them. It typically memorizes only the summary qualities of pitch and timbre. The fascinating element in Helmholtz's theory is that signs, like all symbols, need not be analogous to the things they are signs of. Its weakness and ultimately its untenability is that it still requires that we reassemble the spectral components in order to perceive sounds (a hypothesis that has never been verified or believed plausible, so far as I have been able to ascertain in the scientific literature). Surprisingly, hearing research has forgotten all about the cognitive mystery that Helmholtz's theory of signs was supposed to solve and has since struggled to puzzle together pitch, timbre, and the Fourier spectrogram unsuccessfully. A radically alternative solution might be along the following lines. The neurophysiologic apparatus of the living organism produces motions in response to external acoustic stimuli. These complex motions originate neurophysiologic activities that represent the distribution of the acoustic response originating within the organ of hearing (such the organ of Corti in the cochlea of mammals or in the papilla in birds.) A special spectrogram can be constructed—let us call it 'morphological' as it depends on the anatomical and neurophysiologic configuration of the receptive organ—that visualizes the distribution of the dynamic response generated within the organ, and *not* the functional make-up of the external stimulus in terms of sinusoids or other functions. Such a *morphological spectrogram* captures the epiphonetic landscape better than spectrograms that are rooted in the traditional linguistic and musical categories. Its phenomenological structures will emerge differently in different species as the neurophysiology of the organ of hearing varies accordingly. The acoustic information carried by the incoming sound is spread over the whole range of motions of the organ of hearing. Obviously, there is no particular place along the organ at which the acoustic signal will be encoded as in the Fourier spectrogram. In other words, there is no need for postulating sharp tuning, or sympathetic resonance as a mechanism of biological hearing. Thus, the morphological spectrograms in birds will represent the acoustic scene of the epiphonetic landscape quite differently than in humans. We might be able to run this thought-experiment in bird hearing again in the future, once the neurophysiology and the anatomy of the organ of hearing in singing birds has been better understood and the technologies for computing the morphological spectrograms of sounds as heard by different species have been made available.

Perhaps there is one more mystery whose acoustic origins in the human larynx will be reflected across the epiphonetic landscape. Laughter. We have an intriguing clue. Peals of laughter have Fourier spectrograms that are very weird and uncannily birdsong-like. Might laughter be diffused along the series of biological species? Or might it become physiologically possible for certain different species, and hence be shared by both humans and birds (and by other creatures as well)?

5. Nature's Laughter

Luigi Paolucci was born in Ancona, the capital city of the Marche, the same region where half a century earlier, Italy's most celebrated lyric poet, Giacomo Leopardi, also was born (in nearby Recanati). It is impossible to grasp the stylistic vitality, the philosophical richness, and the scientific originality of Paolucci's essay and his research on birdsong without remembering Leopardi's *Operette morali*, a collection of philosophical meditations, one of which is entitled 'Eulogy of birds'.[74] A comparison between the two works is a paradox. No contrast might be sharper. While Leopardi culled his scientific knowledge of birds from literary sources, mostly from the massive natural history by Georges-Louis Leclerc, Comte de Buffon, Paolucci based his essay on first-hand observation of birds and ecological systems. He learned how to imitate birdsong and carried out experimental research. While Leopardi is a classicist and a literary stylist, who works on words in the closet of his diamond-sharp mind, Paolucci almost tastes biological life in its pristine organismic state, in the nests in the hidden recesses of nocturnal woods landscapes, or at the crack of dawn on the beaches. Yet there are deeper harmonies that resonate between the two characters. The reader of Paolucci's essay will feel—at least I hope so, as this was my original experience—a sense of harmonious happiness throughout the narrative. At multiple junctures Paolucci finds subtle ways of communicating that birds are the incarnation of happiness and that birdsong announces this happiness. This is especially true when he hints at the natural scenarios, in which he carried out certain observations, and conveys the feeling of oneness with the totality of life, a feeling that permeates the ecologies that Paolucci explores and that himself inhabits.

In the *Eulogy of birds* Leopardi put forward a startling argument. Birds are the happiest creatures in the world, not in that we feel happy when we see them (though this may be true as well), but in the sense that they truly feel happiness in themselves. Both their forms and their patterns of behavior demonstrate that they are predisposed to happiness, a phenomenon that cannot be thought of as being an illusion or vain appearance.[75] Listening to birdsong gives us pleasure not so much because of the melodies, or timbre or variety, but because the singing of birds carries the signification of happiness. Birdsong is a sort of laughter, claims Leopardi.[76] It follows, according to him, that birds partake of the human privilege of laughing. Some even think that humans might be defined as *laughing* animals as well as *rational* animals, in that laughter is as proper and as specific to humans as reason.[77] We might conclude, in accord with Leopardi's sentiment, that as reason and laughter reciprocate so birds also partake of human reason. That birdsong is like laughter is an extraordinary claim. Leopardi might have generalized Buffon's serene remark that Mother Nature must have created birds in a mood of hilariousness.[78]

With all the caveats that I already indicated, we must revisit the spectrogram of the nightingale in the light of the concept of epiphonetic landscape, but this time we should look for inspiration that might illuminate Leopardi's intuition. To begin with, I would make the suggestion that we should imagine a finite figure exactly like an equilateral triangle, placing at the two vertexes of the base the vowel sound and the consonant sound, and the laughter sound at the top vertex. We might then regard the triangle as a space for sounds that have distinct, pure qualities at the vertexes, namely, consonant, vowel, or laughter, respectively (let us call them *elemental sounds*), and mixed qualities anywhere else in the middle, which come in variable proportions depending on the location of the representative point. Hence certain types of laughter, for instance, could be regarded as a convergence of the sounds of the vowel and the consonant when they are fused together. In effect, psycholinguists occasionally speak of *voiced* laughter to indicate the idea that among the myriad variations of laughter, we also find patterns that are mixed up with more voice-like sounds. Vice versa, as we move away from the top of the triangle, the elemental laughter looses its pure, elemental quality and begins to merge with vowels and consonants. Anywhere in the middle a sound would have a quality that has different portions of the three elemental sounds. Leopardi suggests that laughter is not original with the human condition, perhaps being only a possibility of expression (made possible, we should add, by the physiology of the larynx and the vocal apparatus), and that in consequence it should rather be considered as an habit acquired by humans at some point in time. While relying on limited ethnographic information, he advanced the hypothesis that laughter may have been discovered by early humans in the state of orgiastic or festive inebriation. Where could we locate the first, inaugural laughter of the inebriated human in the triangle? The epiphonetic landscape must include the infinite variations that human and singing bird laughter produces. Its substance is formed by the elemental sounds. Perhaps we have a hard time picturing in our minds the singing bird as laughing bird, especially because we are used, or have learned to associate laughter with smile and the plasticity of human facial expression. Yet, as laughter remains one of the deepest mysteries of the human condition, it is fascinating to think that birdsong might be a natural pathway to expression and communication that nature excogitated long before the evolution of mammals and which she later resumed as occasion arose in the evolutionary history of life.

This association of laughter and song in both birds and humans that nature experimented with has further philosophical ramifications. In the footsteps of Buffon, Leopardi, and Paolucci, we should note that the two strongest senses in birds are generally sight and hearing. According to Leopardi, this very fact suggests that they may have a highly developed imagination. Their rapid movement and ability to fly over long distances allow them to observe in succession and absorb large amounts of visual and hearing information from multiple viewpoints. We need to make an effort to imagine how much more visionary we should become if we were endowed with the same mobility as birds. Hence their imagination might be very effective in shaping their life-world. But not the imagination that is associated with melancholy and reflectivity, that is, the imagination of the great poets such as Dante or Tasso, or the philosophers (as Leopardi thinks), rather the imagination that is totally spontaneous and innocent, the imagination of childhood, which is about the creative processes of fantasy and nature.[79] This connection between laughter, child-like imagination and the happiness of birdsong brings us back again to the epiphonetic landscape. It might be regarded as summarily exposing a deeper dimension of connectedness in the life-world of organisms, a connectedness that may emerge more articulate under certain evolutionary circumstances and hence turn into a manifestation of the phenomenon of phoneticism, the phonetic languages of humans, or simply remain latent in a state of undifferentiated unity as in birdsong. The idea of epiphonetic landscape and the mystery of laughter offer a viewpoint from which to scan the human/animal continuum in search for quantized points of transition. It is at the intersection of natural history, philosophy, psychoacoustics and empirical research that, in the wake of Leopardi and Paolucci, birdsong should still captivate our imagination and make us innocent again.

Nature's Creative Silence

In Ennead III (8, 4, 1–14) Plotinus has recourse to a paradoxical yet philosophically pregnant prosopopoeia in which *nature* proclaims, through a profusion of rather enigmatic words, that her own nature is but silent creativity by means of theorizing, or contemplation. Silence, we are thus invited to speculate, must crucially contribute to our understanding of the paradoxical discourse of nature. The question that nature supposedly answers through her speech concerns the end of her creative engagement with the world of forms, namely, the *that-for-the-sake-of-which* of her theorizing activity. The paradoxical aspect of the discourse is reinforced by nature's assertion that she engages in theorematic contemplation, in the style of the geometers who draw figures. However, she does not draw linear diagrams but letters or profiles of bodies.[80] In Leopardi's *Operette morali*, a work permeated with Plotinian figures and visions throughout, a similar prosopopoeia is presented in the *Dialogue of nature and an Icelander*. Here a traveler from Iceland, who has desperately been fleeing nature all his life, reaches the torrid zone of central Africa where a gigantic female figure, beautiful and tremendous at the same time, sits in a resting mood, reclining over a mountain. When pressed on by the unexpected visitor, she reveals that she is nature in person and 'answers' the Plotinian question regarding the *that-for-the-sake-of-which* of her theorizing activity—which, according to Leopardi, results in the production of biological life in the universe and of consequent, irredeemable human suffering—by remaining dead silent. At which point, the exhausted traveler is devoured by two hungry lions. I will follow Leopardi's vision in trying to decipher the enigmatic discourse of nature in *Ennead* III (8, 4) and the *Dialogue*, by making the suggestion that a metaphysical silence, of which its acoustic counterpart is but an impoverished instantiation in the world of sensible appearances, is the ineliminable hence paradoxical precondition of her creative contemplation, through which she engages the world of forms by means of theorematic, or contemplative activity. The geometer's diagrams look like the hieroglyphs that the ancient sages of Egypt employed in their scriptures (Ennead V, 8, 6), but the latter are more autonomous than the diagrams of geometry since geometric proofs always need to be supplemented with a textual narrative which must guide the geometer towards the conscious achievement of proof. Nature is forever captivated in a sleepy, semi-conscious or perhaps unconscious state, in which she gives birth to embodied hieroglyphs (living organisms that are in need of no proof) when she follows the patterns of her creative, silent contemplation. The logic of life, Plotinus seems to be saying, is in no need of proof, it is pure existence, but at the price of a poverty of consciousness, a sleepiness, a fateful remaining silent about the ultimate question, the dilemma concerning the *that-for-the-sake-of-which*.

The personification of nature appears promptly at a crucial juncture in the text when Plotinus raises the fundamental question of the reason for her creative engagement with forms. What is the end of nature's contemplative creation? She addresses the questioner by saying that the question should not have been asked in the first place, and that the questioner should rather have made an effort to understand in silence, doing what she does, namely, remaining in silence. What is there to understand, after all, urges nature? It is this. That what she generates is the object of her contemplation, resulting from silence, that it is her very nature to generate in silence, and finally that she is endowed with this generative power by contemplation because she herself is the daughter of a higher contemplation. Then the metaphor is invoked of the geometrical style of contemplative creation by drawing pictures or diagrams. The metaphor, I wish to suggest, illuminates the concluding part of the discourse, in which nature returns to the image of her own genesis from her parents by a contemplation which needed no acting, for all that was needed, she insists, was that her parents were reasons of a superior order contemplating (hence creating) themselves. A geometrical proposition, for instance, the so-called *Pythagorean* theorem (from the first Book of Euclid's *Elements*), is typically demonstrated through a process of synthesis in which a complex figure, a diagram, is drawn first. It is the visual ground for a text that will walk the reader through the proof. But the narrative which is weaved around the diagram will necessarily make a number of detours by invoking other propositions and ultimately indemonstrable first principles, on which the entire deductive structure of geometry is founded. The diagram is indispensable, one must not forget this. It is the diagram that takes center stage insofar as certain intermediate conclusions that prepare the way for clinching the final statement are based on information that emerges in the diagram, and which is not available in the text. This is not due to a stylistic preference of the author, but precisely to the nature of Greek geometry, which is squarely diagrammatic hence contemplative, and which cannot be reduced to symbolic manipulation in terms of, say, algebraic equations. However, the text remains indispensable too. One cannot entirely get rid of it. Geometry cannot be performed silently. This conclusion holds true literally, as can be gathered from recent historiography of Greek geometry, which has highlighted the vociferous social dimension in which early geometers operated. Nature is more powerful than geometry in that she creates by giving birth to living organisms that are in need of no supplementary commentary in order to accomplish their own existence. Organisms are self-proving. The power of life comes from silence, from a certain necessity that need not be articulated linguistically, or logically. Nature, we should like to imagine, would be able to contemplate the truth of the Pythagorean theorem, theorizing all the way through its proof, as far as reaching the first principles and without needing a text that reminds her of the already demonstrated propositions and indemonstrable first principles (on which the former ultimately rest), a praxis that requires some time to be articulated. This pragmatic requirement weakens the contemplation of the geometers, who are left with not choice but engage in noisy discourse. Plotinus comments that even though we might want to grant nature a form of knowledge or perception, these would not resemble the faculties of other beings. It would be like comparing the faculties of somebody who is at sleep with those of someone who is awake. Nature remains in a sleepy mood while contemplating her being in herself and with herself and her identifying herself with the object of her contemplation. It is a silent contemplation, a little obscure, as Plotinus admits, and thus nature's creatures are somewhat weak ones. This happens precisely because a weak contemplation produces a weak object of contemplation. At this point we must ask ourselves: Has nature really answered the inquirer's question concerning the *that-for-the-sake-of-which* of her creative theorizing? What is the end of nature's contemplative creation? Though nature has for a moment violated her discipline of silence, she has not revealed herself.

When the Icelander reaches the tropical area of central Africa, after fleeing from Iceland and roaming the world in search of a safe haven from the destructive power of nature, he is astonished to find there his very enemy, nature herself, in the guise of a gigantic woman. Her traits were inspired to Leopardi by the early reportages and intriguing iconography of the strange and enigmatic statues found by European colonialists on Easter Island. At first nature is surprised to see the traveler, as she notices that the human species has until now not settled in those remote areas, where her powers, she warns him, are the most annihilating. Undeterred, the colonialist challenges nature, by launching into an accusatory peroration that lists all the incessant evils that she visits upon the human race, such as weather, earthquakes, natural disasters, and above all the very fact that she inscribes the inexorability of ageing, illness, and death into the very reason of life itself. Her first line of defense is the following. She claims that in all her operations, with few exceptions, she does not care for human happiness or unhappiness, and that when she happens to injure us in whatever way, she does so while almost always unaware of it. Even if she were to extinguish the entire human species, she reiterates that she would not be aware of it. The Icelander replies that if nature has placed humans and all living organisms in the world she is responsible for their care, or at least for making sure that they should live a decent life, given that life did not originate by its own accord against the will of nature. Nature's second line of defense is an ecological argument. The whole universe is an organized whole in which the cycle of generation and corruption is instrumental to the preservation of the whole. Therefore, if there were anything in the universe that did not suffer annihilation, the whole would suffer. The Icelander quickly retorts that this is the hackneyed refrain repeated in more or less the same vein by all philosophers. He presses on insisting that the fundamental question has been left unanswered by nature: Tell me what no philosopher has ever been able to say, namely, who likes or who benefits from this most unhappy life of the universe, which is preserved by the suffering and death of all the things that make it up? At this moment, two lions, black and blue by starvation, put an end to the niceties of philosophical gallantry. They argue by devouring the Icelander.[81]

Elsewhere, in a series of annotation to his collected poems, or rather, as I prefer to call them, prayers, Leopardi puts forward a startling insight about the opposition of ancients and moderns. He claims that while nature spoke to the ancients, she did not reveal herself to them.[82] Conversely, she has revealed herself to us, the moderns, but she no longer speaks to us.[83] What Leopardi intimates is that modernity has brought us science, whose task it is to find out about the truth, and indeed, according to Leopardi, science has carried out this assignment with great success over the last two centuries (four, if we count at the beginning of the twenty-first, and if we agree with him). But he also points out that truth is a coincidence of opposites, and indeed in one of his poems he speaks of the "black face of truth", a mind-boggling paradox which flies in the face of millenary metaphors of light drifting away in the Western discourses about truth.[84] Thoroughly acquainted with the religious, poetic, and philosophical traditions of the Greeks and with the Greek language, Leopardi seems to imply that Plotinus's nature spoke to him in a certain pristine way, while however not revealing herself, not showing her true emptiness. She could not answer the ultimate question of the *that-for-the-sake-of-which*, which was raised by the ancient philosopher. But if truth is itself a coincidence of opposites, how did nature eventually manage to reveal herself to us, and why has the compensatory effect of this revelation resulted in her withdrawing from linguistic intercourse with humans? Modernity is the locus of the great philosophical systems, the orderly systems of symbols in which propositions about the real are formed in order to organize a coherent, encyclopedic structure of truths materially inscribed in books, paper, and ink—a beautiful philosophical figure explored in depths by Daniel Selcer.[85] Science is one of these symbolic systems, a body of knowledge polarized by the metaphors of materialism, mechanism, and rationalism. But all that science has discovered, Leopardi charges, is the nullity of the universe, and hence the nullity of the idea of truth, forever captivated in a vicious circle of endless linguistic self-references. The universe is pure existence but it comes about at the price of a poverty of consciousness, a sleepiness in which we, as part of nature, are captivated, the children of a weakening of contemplation that emasculates us.

The fascinating figure of the sleepiness of nature, her creative contemplation in that mercurial, liminal zone separating wakefulness from sleep, was evoked by Plotinus and followed through by Leopardi. It invites resistance to the rationalizations of modern psychologies of the unconscious, by which I mean their tendency to speak of the unconscious by falling back upon metaphors from ordinary conscious discourse. At the threshold of European Romanticism and the turning of the century of the unconscious, which ended with Freud's *Interpretation of dreams*, Leopardi offers insights beyond the strictures of rationalization (to which all too often Freud and sometimes Jung capitulated), thus demonstrating possibilities of poetic speech that rationalistic modernity and postmodernity have otherwise condemned to oblivion. To 'speak' about the ultimate question of the *that-for-the-sake-of-which* of creative contemplation means a radical silence, on which Plotinus inquired masterfully, a philosophical or metaphysical silence, in the sense of a letting go of rational language, an escape from the labyrinth of argumentative consciousness, which loses itself in the perfunctory symbolisms of modernity. Instead, 'speaking' about the ultimate question requires a felicitous resorting to poetic language, not in order to recover novel conceits from the ornaments of poetic images, but because poetry dramatically pushes us through what Mircea Eliade called 'a rupture of planes'.[86] Enticing us to move away from the ontology of being as either presence at hand or readiness to hand, towards an ontology of prayer, Leopardi, looking backwards towards Plotinus, invokes the idea of truth as a coincidence of opposites, an open enterprise, an opus that might redeem us and by so doing entice nature to speak to us again. An idea of truth, for which truth itself "blurs into a limit-experience of light, or, if you like, within a state of eternal, luminous absence, within a radiant emptiness; [in which 'emptiness'] is being that has overcome all its properties, or rather a non-being, utterly positive, which offers an immaterial happiness, without a substratum, without any foundation in any world".[87]

Nature's Creative Laughter

"Le néant pour le bouddhisme (à vrai dire pour l'Orient en général) ne comporte pas la signification quelque peu sinistre que nous lui attribuons. Il se confond avec une expérience-limite de la lumière, ou, si on veut, avec un état d'éternelle absence lumineuse, de vide rayonnant c'est l'être qui a triomphé de toutes ses propriétés, ou plutôt un non-être suprêmement positif qui dispense un bonheur sans matière, sans substrat, sans aucun appui dans quelque monde que ce soit".[88] And perhaps, by way of magical enchantment, ecstatically or unconsciously, both Leopardi and Paolucci have followed the light of nature—the *lux naturae* of the spiritual alchemists, an arcane light revealing herself both in silence and in human and non-human song—along their apparently diverging pathways towards the rarefied experience of truth symbolized in the smile of the Buddha. This premodern pathway to truth that Paolucci explored in his quest for the archaic truth of birdsong matured in an ecological life-world still pristine as an experience of human dwelling among non-human natural things. This same premodern pathway was elucidated by Francis Bacon in a phantasmagorical, superbly illuminating metaphor, and it is certainly bewildering that long-standing, diehard positivist readings of Bacon that are by and large still current in both the academic and the pulp literature—not to mention the canonic narratives of the scientific revolution of the seventeenth century dolloped out everyday in popular classes to innocent undergraduates—have practically obliterated (even in post-modern and post-human consciousness) this premodern, non-mechanistic, spiritualist, or rather anarchist dimension that the Lord Chancellor opened up in his *Novum organum*. "But we, by long dwelling among the things of nature in chastity, have not abstracted our understanding further away from natural things than the exact distance required for their luminous images to come into focus, so that little, if anything, has remained for reason to brag about her powers. Indeed, in both learning and teaching we bowed to the same humility".[89] This juvenile dwelling among the things of nature in absolute *chastity* will be sensed by the reader throughout Paolucci's essay. It is a background, rhythmic pulsation, an ethical echo that resonates with clarity in his writing. It is the ethical force of the modality of this dwelling that Bacon flagged in his luciferous Latin with the adverb 'caste' [chastely]. Purity, child-likeness, unswerving attentiveness to the finest sonorities of the things of nature, in humble prayer, hoping for them to emerge in the clarity of the images that are brought into focus by dint of self-sacrificing perseverance and after making peace with the silences of nature with which her score is ruptured throughout. This mode of life as chaste pursuance of the light of nature that finally reveals herself in the smile of birdsong is a renunciation, a giving-up of the violent assault on the world that was spearheaded by the irresistible ascendance of the modern ego, and which is proclaimed by Bacon at the same time in the same book. It is an exquisite example of the dissonant music that premodern philosophy, uncontaminated by the clownish absurdity of the *cogito-ergo-sum*, weaved in its heraldic works. It is not farfetched to institute a simile between the smiling Buddha and this ecological mode of dwelling among the beings of nature in chastity, a mode that cries for resuscitation or rather great instauration against the toxic and *de-human/animalizing* idolatries of the postindustrial, globalizing age, against the propaganda that glorifies the destruction of nature's unique singularities. Both are a manifestation of the peace that renouncing the histrionics of the Fascist ego and its neuroscientific correlates brings to the *visage* of universal existence. For, as the readers will perhaps experience (if they make allowances for the poverty of the translator's wits), Paolucci's language, though disguised as scientific language and paying the occasional homage to the bastard spirit of our time, betrays the deeper existential and *ethical* ingenuity of the ego-less dweller among beings of nature, a benign resident of the ecologies of birdsong, who has become himself empty of arrogance, a non-violent hunter, and whom nature has rewarded by inebriating him with the sound of cosmic laughter. The sound of perennial creation.

Notes

1. Levinas 1991, pp. 109–110. "Is not the evil of suffering—extreme passivity, helplessness, abandonment and solitude—also the unassumable, whence the possibility of a half opening, and, more precisely, the half opening that a moan, a cry, a groan or a sigh slips through—the original call for aid, for curative help, help from the other me whose alterity, whose exteriority promises salvation?" (Levinas 1998a, p. 93). Levinas's text raises an interesting question as to the interpretation of the word 'couverture' (literally 'covering', or 'covering up') and hence of the whole passage. The context seems, at first glance, to require precisely the opposite meaning, namely, that of an opening or at least that of a permeability. While the English translators have opted for rendering the ambiguous 'couverture' as 'half opening', which makes the sentence flow better (even though they miss an important portion of Levinas's thought by inadvertently ignoring the subordinate sentence "...et, ainsi, de par sa non intégration dans l'unité d'un ordre et d'un sens"), the Italian version that I have consulted and generally relied upon for a comparative reading has retained the ambiguity and rendered the original as 'copertura' (Levinas 1998b, pp. 125–126). The 'couverture' per se, or possibly the evil of suffering, extreme passivity, that cannot be normalized in the pre-existing unity of an order or a meaning (that is missing in the English rendition), and hence through which a cry or a moan or a groan or a sigh can still slip through, could then be interpreted as a last layer in the permeable stratification that protects the sufferer from the ultimately absurd suffering of appealing to the other human being for their therapeutic help—an appealing that would expose being defenseless, hopeless, that would force one confess one's naked powerlessness. I tend to prefer the latter, not least because it speaks deeply to my personal experience of suffering, where my effort to make an appeal to the other brings about an extreme level of pain. But there could be just a typo in the intended original phrase, which, in Levinas's mind, should have had 'ouverture'.
2. The chapter "Useless suffering" in Levinas's book, 'Entre nous' (1991), had been published as a paper in 1982, in the Italian journal, *Giornale di metafisica* (Levinas 1982), together with a number of contributions by other philosophers on the theme of suffering.
3. Levi 1991 [1986], pp. 83–101. Cf. the reviews of, and commentaries on, Levi's book (and a reply from him, "The black hole", written three month before he took his life), in the 1991 edition, pp. 169–181.
4. Levi 1991, p. 89.
5. "Mi pare significativo ricordare questi abomini in un'epoca in cui, con ragione, viene messo in discussione entro quali limiti sia lecito condurre esperimenti scientifici dolorosi sugli animali da laboratorio. Questa crudeltà tipica e senza scopo apparente, ma altamente

simbolica, si estendeva, appunto perché simbolica, alle spoglie umane dopo la morte..." (Levi 1991, pp. 99–100). ["I think that it is important to recall these abominations at a time when, quite rightly, the question arises whether and to what extent it is ethical to carry out painful scientific experiments on laboratory animals. This typical cruelty, apparently without purposes, yet highly symbolic, was extended, precisely because symbolic, to the human remains after death..."]. It is to be noted that the moment of rupture in consciousness caused by the surfacing of the repressed content, i.e., the non human-animal suffering, is immediately alleviated, and the symptom of silence quickly disappears. Levi continues the narrative by returning to the theme of the symbolism of useless violence in relation to the abominable uses of human remains made by the Nazis, particularly the 'ingenious' transformation of cadavers and parts of human bodies into construction materials.

6. Cf. the *Vocabolario etimologico della lingua italiana* (Pianigiani 1907, ad voces, *vultus* and *volere*).
7. Cicero 1853, p. 410 (slightly altered). "Ipsum autem hominem eadem natura non solum celeritate mentis ornavit sed et sensus tamquam satellites attribuit ac nuncios et rerum plurimarum obscuras nec satis... intellegentias incohavit, quasi fundamenta quaedam scientiae, figuramque corporis habilem et aptam ingenio humano dedit. Nam cum caeteras animantes abiecisset ad pastum, solum hominem erexit et ad caeli quasi cognationis domiciliique pristini conspectum excitauit, tum speciem ita formauit oris, ut in ea penitus reconditos mores effingeret. Nam et oculi nimis argute quem ad modum animo affecti simus, loquuntur et is qui appellatur vultus, qui nullo in animante esse praeter hominem potest, indicat mores, cuius vim Graeci norunt, nomen omnino non habent" (Cicero 1872, p. 350).
8. "Sed Graecis cum vocabulum non desit, vis tamen notationis deest, in qua inest indicium voluntatis" (Cf. Turnèbe's commentary, in Cicero 1552, p. 28)
9. Cf. his *Vocabolario etimologico della lingua italiana* (Pianigiani 1907).
10. "Le visage ne se ressemble point à la forme plastique, toujours déjà désertée, trahie par l'être qu'elle révèle, comme le marbre dont, déjà, les dieux qu'il manifeste, s'absentent. Elle diffère de la face animale où l'être ne se rejoint pas encore dans sa stupidité de brute." (Levinas 1957, p. 248). The sentence might sound ambiguous as the pronoun 'elle' cannot refer to 'visage'. However, the meaning seems to require it. My feeling is that 'elle' refers to the noun *épiphanie* in the previous sentence where the *visage* is defined as epiphany (feminine in French), hence my solution in the translation of this text. A second difficulty in the text (to my mind) is the use of the reflexive verb 'se rejoindre' evoking multiple meanings such as meet, meet up, join, be similar, or encounter.
11. Derrida 2008, p. 106.
12. "There is, to my knowledge, no attention ever seriously given to the animal gaze, no more than to the difference among animals, as though I could no more be looked at by a cat, dog, monkey, or horse, than by a snake or some blind protozoon" (Derrida 2008, p. 107). Yet this is not the case, as we shall see presently, for Renaissance physiognomy, for example, was definitely and very seriously engaged with the animal gaze. I find Derrida's critique of Levinas's notion of *visage* cogent, but he seems to have little to offer towards a renewal of philosophy and ethics that might overcome the humanist prejudice inherent in both his own and Levinas's thought.
13. Levinas 1957, pp. 241–245.

14. Inspired by Plutarch's celebrated dialogue (*De facie quae in orbe lunae apparet*, or *Concerning the face which appears in the orb of the moon*), what follows in this section is loosely based on the text, comments and translation given by Harold Cherniss, in Plutarch 1957, pp. 1–223. The Latin word *facies* in the title translates the Greek *prosōpon*. The latter returns periodically at key junctures in Plutarch's Greek text.
15. Bernabò Brea 1998.
16. Indeed, at the very beginning of the dialogue, we are warned against easy solutions. "As people with chronic diseases when they have despaired of ordinary remedies and customary regimens turn to expiations and amulets and dreams, just so in obscure and perplexing speculations, when the ordinary and reputable and customary accounts are not persuasive, it is necessary to try those that are more out of the way and not scorn them but literally to chant over ourselves the charms of the ancients and use every means to bring the truth to test" (Plutarch 1957, p. 35). The archaic revivalism of "chant[ing] over ourselves the charms of the ancients" evokes a phantasmagoric return to a lost value system that harkens back to a prehistoric epoch when truth-testing was not a matter of verification of hypotheses but rather a recourse to the resources of cosmic life itself as the fount of pristine knowledge. The pristine knowledge that doubtless was the preserve of pre-agriculturalist, shamanistic cultures.
17. Plutarch 1957, p. 141.
18. Plutarch 1957, p. 141. Emphasis is mine. Compare and contrast this passage with Galileo's fabulous description of his seminal telescopic observations of the face of the moon. "Let me speak first of that surface of the moon which is turned towards us. For the sake of being understood more easily, I distinguish two parts in it, which I call respectively the brighter and the darker. The brighter part seems to surround and pervade the whole hemisphere; but the darker part, like a sort of cloud, discolors the moon's surface and makes it appear covered with spots. Now those spots, as they are somewhat dark and of considerable size, are plain to everyone, and every age has seen them, wherefore I shall call them great or ancient spots, to distinguish them from other spots, smaller in size, but so thickly scattered that they sprinkle the whole surface of the Moon, but especially the brighter portion of it. These spots have never been observed by anyone before me; and from my observations of them, often repeated, I have been led to that opinion which I have expressed, namely, that I feel sure that the surface of the Moon is not perfectly smooth, free from inequalities and exactly spherical, as a large school of philosophers considers with regard to the Moon and the other heavenly bodies, but that, on the contrary, it is full of inequalities, uneven, full of hollows and protuberances, just like the surface of the Earth itself, which is varied everywhere by lofty mountains and deep valleys" (Galilei 1880, p. 15). Rationalist interpretations of Galileo, in the wake of nineteenth-century positivism and of its logical mutant in the twentieth, have presented him as the inaugurator of modern telescopic astronomy, which is certainly true. But they have completely missed the magical-hermetic culture that had primed his mind, and which disrupts the narratives of both historians and sociologists of science. I take the liberty of referring the reader to Palmieri 2016, for a presentation of Galileo's hermeticism, in the spirit of continental phenomenology, which is a radical departure from the strictures of both positivistic historiography and the philosophical confusion of analytic sociology of science.

19. "After all, in what sense is earth situated in the middle and in the middle of what? The sum of things is infinite; and the infinite, having neither beginning nor limit, cannot properly have a middle, for the middle is a kind of limit too but infinity is a negation of limits. He who asserts that the earth is in the middle not of the sum of things but of the cosmos is naïve if he supposes that the cosmos itself is not also involved in the same difficulties. In fact, in the sum of things no middle has been left for the cosmos either, but it is without hearth and habitation, 80 moving in infinite void to nothing of its own; or, if it has come to rest because it has found some other reason for abiding, not because of the nature of its location, similar inferences are permissible in the cases of both earth and moon, that the former is stationary here and the latter is in motion there by reason of a different soul or nature rather than a difference of location." (Plutarch 1957, p. 77).
20. Plutarch 1957, pp. 83–85 (slightly altered).
21. Plutarch 1957, p. 89.
22. "Such is probably the case with the cosmos too, if it really is a living being: in many places it has earth and in many fire and water and breath as the result not of forcible expulsion but of rational arrangement. After all, the eye has its present position in the body not because it was extruded thither as a result of its lightness, and the heart is in the chest not because its heaviness has caused it to slip and fall thither but because it was better that each of them should be so located. Let us not then believe with regard to the parts of the cosmos that earth is situated here because its weight has caused it to subside or that the sun [...] was extruded into the upper region like an inflated skin by reason of its lightness or that the other stars got into their present positions because they tipped the balance, as it were, at different weights. On the contrary, the rational principle is in control; and that is why the stars revolve fixed like 'radiant eyes' in the *visage* of the cosmos, the sun in the heart's capacity transmits and disperses out of himself heat and light as if it were blood and breath, and earth and sea 'naturally' serve the cosmos to the ends that bowels and bladder do an animal. The moon, situate between sun and earth as the liver or another of the soft viscera is between heart and bowels, transmits hither the warmth from above and sends upwards the exhalations from our region, refining them in herself by a kind of concoction and purification." (Plutarch 1957, pp. 91–93, slightly altered, emphasis mine). An extraordinary passage, visionary and lucid, this recognition of the cosmos as a living organism palpitates with the fascination of the face. The radiant eyes in the *visage* (Greek *prosōpon*) of the universe have an unmistakable psychedelic brilliance. One wonders whether in the melting pot of Greco-Roman culture at the time when Plutarch wrote his essay (second half of the first century AD), psychotropic foods (possibly of Middle Eastern and/or North-African origin) might have offered the experience for his characters's visionary tales (unless it was the experience of the Eleusinian Mysteries).
23. Levinas 1957, p. 249.
24. "Que le mouvement de l'âme qui est plus que cognitif que la connaissance, puisse avoir une structure différente de la contemplation—voilà le point sur le quel nous nous séparons de la lettre du Cartésianisme." (Levinas 1957, p. 249).
25. Plutarch 1957, pp. 197–199.
26. Plutarch 1957, pp. 215–217. Emphasis mine.

27. The myth is recounted by the Carthaginian Sulla, who learned it from a stranger who had come to visit Carthage from a mysterious island, where in turn he had been told the myth by daemons serving Cronus. The latter is shaken by powerful emotional waves as he sees in dreams what Zeus premeditated. "Cronus himself sleeps confined in a deep cave of rock that shines like gold—the sleep that Zeus has contrived like a bond for him—and birds flying in over the summit of the rock bring ambrosia to him, and all the island is suffused with fragrance scattered from the rock as from a fountain; and those spirits mentioned before tend and serve Cronus, having been his comrades what time he ruled as king over gods and men. Many things they do foretell of themselves, for they are oracular; but the prophecies that are greatest and of the greatest matters they come down and report as dreams of Cronus, for all that Zeus premeditates Cronus sees in his dreams and the titanic affections and motions of his soul make him rigidly tense until sleep restores his repose once more and the royal and divine element is all by itself, pure and unalloyed." (Plutarch 1957, p. 189). The eschatological myth concludes the dialogue. It is the fusion of the ethnic styles of the imagination that are incarnated in the main characters. The oracular power of these demonic servants of Cronus is the function of certain souls that have been separated from the terrestrial body on the moon already, but which have lingered and returned to earth as spirits. The insistence on the dreaming state of Cronus should be related to the moon and its soul substance. The consciousness that arises in soul is moon-like, dreamy, intermediate between the material terrestrial body which the soul enfolds and keeps at bay (thus preventing its exaggerated egotistic expansion (as in modem philosophy) and the solar intellect in which all vestiges of subjectivity will be dissipated.
28. Nardi 1967, pp. 3–39.
29. Both the text and the excellent translation are accessible at the *Digital Dante* project of Columbia University: https://digitaldante.columbia.edu/dante/divine-comedy/paradiso/paradiso-2/. Cf. Julie Van Peteghem. "What is Intertextual Dante?" Intertextual Dante, Digital Dante. New York, NY: Columbia University Libraries, 2017. https://digitaldante.columbia.edu/intertexual-dante/.
30. Generations of Italian students, including the author, have been treated to this nonsense by idiotic ministerial programs, autocratically enforced by governments both Fascist and Republican. But there is a deeper issue that has badly affected Dante scholarship especially in the twentieth century. It is the *normalization* of Dante as a national poet of the newly born Italian state, a process which has obscured and exorcised the esoteric and magical dimensions of the man and the poet by denigrating interpretive work that went against the positivist and political orthodoxy and instead pointed at the undercurrents of Dante's visionary thinking. Cf., for instance, Pascoli 1902, an inspiring work that never made it into mainstream educational programs.
31. From the *University of Toronto Libraries collection* of Internet Archive at https://archive.org/. Freely available at: https://archive.org/details/ladivinacomm192102dantuoft.
32. Text and translation from the Digital Dante project of Columbia University: https://digitaldante.columbia.edu/dante/divine-comedy/paradiso/paradiso-4/.
33. There are other spirits that share Piccarda's fate. All have been complicit in giving in to coercion. But their uniqueness remains intact, a different incarnation of the same mystery, the freedom of the will. The coincidence of opposites is also the paradox of the uniqueness

of every existence despite its being a manifestation of the same mystery. Beyond the dualism autonomy/heteronomy that pervades modern philosophy, the paradox of the coincidence of opposites indicates a pathway to thinking otherwise than human.

34. "Quest'arte dimostra all'improvviso dall' aspetto del volto, senz'altra esperienza, come celeste oracolo e divina arte, o qual si voglia altro velocissimo modo d'indovinare, quai scelerati scacciare e quai onorati abbracciar devi" (Della Porta 2013, p. 2).
35. In a disclaimer at the end of the dedication of the first Latin edition, Della Porta wrote "Haec scientia coniecturalis est, nec semper optatum assequitur finem: cuius signa naturales tantum propensiones indicare possunt, non autem actiones nostrae liberae voluntatis, vel quae ex vitioso, vel studioso habitu dependent: nam in bonis, malisque actionibus, quae in nostra potestate sunt, virtus, et vitium consistunt, non autem in propensionibus, quae in nostra voluntate non sunt" (Della Porta 1586, fourth page, not numbered). The book obtained the imprimatur, which was printed at the end of the book.
36. Della Porta 2013, p. 122.
37. "...coloro che vogliono far profitto in questa scienza, bisogna che studino con grandissima diligenza I libri delle istorie degli animali, acciò siano bene informati de' costumi, de' pensieri, delle descrizioni et insiememente delle lor parti; perchè da quei depende tutta la scienza" (Della Porta 2013, p. 46).
38. Della Porta 2013, pp. 481–482.
39. Della Porta 2013, p. 480.
40. Della Porta 2013, pp. 557–587.
41. What follows in my report is taken especially from Flourens 1842, a massive treatise in which the French scientist reports and discusses in great detail his researches on the brain and brain function in vertebrates. Cf. Legée 1992, a collection of studies with ample documentation and formerly unpublished primary sources.
42. Flourens 1842, pp. 37–43.
43. Bossuet 1863, pp. 155–195.
44. Delamarre and Brunetière 1907.
45. Bossuet 1863, pp. 171 ff.
46. Bossuet 1863, pp. 181–182.
47. See the many interesting essays edited recently in Boulogne 2005, with vast bibliographies.
48. Foucault 1999 [1983]. Foucault gave six remarkable lectures on parrhesia at the University of California at Berkeley in 1983. "The text was compiled from tape-recordings made of six lectures delivered, in English, by Michel Foucault at the University of California at Berkeley in the Fall Term of 1983. The lectures were given as part of Foucault's seminar, entitled 'Discourse and Truth'. Since Foucault did not write, correct, or edit any part of the text which follows [on the website], it lacks his imprimatur and does not present his own lecture notes. What is given here constitutes only the notes of one of his auditors. Although the present text is primarily a verbatim transcription of the lectures, repetitive sentences or phrases have been eliminated, responses to questions have been incorporated, whenever possible, into the lectures themselves, and numerous sentences have been revised, all in the hope of producing a more readable set of notes." Cf. the editor's note by Joseph Pearson, at: https://foucault.info/doc/documents/parrhesia/index-html. The revision is occasionally arbitrary, in my opinion, even though in fairness to the editor it must be said that in his

exposition Foucault does not always seem to be clear and consistent (at least to me). Fortunately, the original tape recordings of the seminar can be listened to at: http://www.lib.berkeley.edu/MRC/foucault/parrhesia.html. It is fascinating to hear Foucault struggle with the English language, with impossible Greek names, and a bunch of monotonous if not self-conceited students, a telling example of the decadence and futility of academic education in industrialized societies.

49. Foucault 1999 [1983].
50. Foucault 1999 [1983].
51. Plutarch 1957, p. 504, in the Loeb edition with facing Greek.
52. Plutarch 1957, p. 505.
53. Plutarch 1957, p. 505.
54. "I should note that I never found any texts in ancient Greek culture where the parrhesiastes seems to have any doubts about his own possession of the truth. And indeed, that is the difference between the Cartesian problem and the parrhesiastic attitude. For before Descartes obtains indubitable clear and distinct evidence, he is not certain that what he believes is, in fact, true. In the Greek conception of parrhesia, however, there does not seem to be a problem about the acquisition of the truth since such truth-having is guaranteed by the possession of certain moral qualities: when someone has certain moral qualities, then that is the proof that he has access to truth—and vice-versa. The 'parrhesiastic game' presupposes that the parrhesiastes is someone who has the moral qualities which are required, first, to know the truth, and secondly, to convey such truth to others" (Foucault 1999 [1983]).
55. The implicit Darwinism in Helmbold's approach to Gryllus is made explicit by his rendering of the Greek in the following passage. "The point that best proves this, gentle Odysseus, is the fact that in beasts valor is natural equal in both sexes and the female is in no way inferior to the male. She takes her part both in the *struggle for existence* and in the defense of her brood" (Plutarch 1957, p. 505, at 987F, emphasis mine). It is anachronistic, in my view, to read Gryllus's parrhesia as a Darwinian manifesto *ante litteram*.
56. Plutarch 1957, p. 503.
57. Plutarch 1829, p. 486. Note the different choice of tenses for the crucial opening statement. 'Tu vedi', here in the present, as opposed to Helmbold's future in the parenthetic 'you will observe'.
58. Stephen Newmyer has aptly noted that Plutarch's support for animal reason goes beyond the confines of logical argument. "Plutarch's case makes an appeal to his proof for rationality in animals to the extent of arguing that creatures demonstrating traces of this rationality cannot be excluded from human moral consideration, but he advances beyond the confines of a reasoned-based argument to include the demands of fellow-feeling for other species that can be shown to suffer as do their human counterparts" (Newmyer 2006, p. 65).
59. See Eliade 1964.
60. The video is available at: https://www.youtube.com/watch?v=P5F_WW6uYhY. This famous song was inspired to Dalla by an anecdotal love story about the tenor Caruso shortly before his death.
61. Nettl 1974, p. 6. Also in the Western classical musical tradition, as Nettl suggests, we can see that the polarity composition/ improvisation is questionable, "by comparing, for example, the painstaking and often protracted method of Beethoven with Schubert's quick, spontaneous creation of lieder" (ibid.).

62. Sawyer 1996, p. 273.
63. Sawyer 1992, 1996, p. 279.
64. Helmholtz 1863, 1877, 1885.
65. Corti 1851, p. 163.
66. It was Georg Ohm's achievement to have shown that the fluid mechanics of the simple one-disk siren could be investigated by means of Fourier analysis. According to Ohm, the sound of the siren could be mathematically represented as a Fourier series of pure tones. "The term 'Ohm's law' traditionally denotes the formula of Georg Simon Ohm relating voltage, current, and resistance in metallic conductors. But to students of sensory physiology and its history, 'Ohm's law' also denotes another relationship: the fundamental principle of auditory perception that Ohm announced in 1843" (Turner 1977, p. 1).
67. Helmholtz 1885, p. 145.
68. I am saying *generally* because it is clear that sound perception is largely a subjective phenomenon. I conducted qualitative experiments in hearing during a seminar with graduate students, some of whom had received some sort of exposure to music and/or some Western classical music training, while other had not had any such experience. For the latter group the concepts of pitch and timbre did not make much sense. It became clear in conversation that salient qualities such as pitch, which seem so 'obvious' to students of music, are not at all obvious to persons who have never learned the language of Western music theory.
69. Cf. Haar 1910, a technical article in which Haar demonstrates that his set of functions meets the orthogonality (and other more precise) requirements that are needed for it to work as a system of functions that can represent more complex periodic functions (in fact any periodic function that satisfies minimum criteria for eligibility).
70. Gleich and Langemann 2011, p. 81.
71. "Cross-sections of avian basilar papillae, at first glance, do not show any of the prominent and uniquely mammalian features, such as the tunnel of Corti or the strict arrangement of one row of inner and three rows of outer hair cells… However, looking more closely at the fine structure of the hair cells and especially their innervation, a very salient analogy emerges. Like mammals, birds have specialized their hair cells into two types, here called tall and short hair cells. Unlike in mammals, those types are not sharply separated" (Köppl 2011, p. 66). This partial and tantalizing analogy between the hair cells distributed in the inner ear papillae of birds and in the cochleas of mammals is highly important in that the functional separation of the two types has led to the recognition that one type of cells does not work as receptor unit but rather as a power unit that, as some researchers think, amplifies the overall receptivity of the organ. In other words, both in birds and mammals, hearing is not a passive neurophysiologic system but an active one. Crucially, the fact that hearing is active implies that the acoustic world can be rendered quite differently in different species according to how the stimuli are 'reinforced' by the apparatus.
72. The segment is from the following online recording available at: https://www.youtube.com/watch?v=NK2_bcQcoD4 .
73. Jablonca and Lamb 2005, p. 1.
74. Leopardi 1895, pp. 270–284. I will quote original excerpts (the lexical and syntactic nuances of which are almost untranslatable) and paraphrase the texts loosely.

75. "Sono gli uccelli naturalmente le più liete creature del mondo. Non dico ciò in quanto se tu li vedi o gli odi, sempre ti rallegrano; ma intendo di essi medesimi in sé, volendo dire che sentono giocondità e letizia più che alcuno altro animale… Gli uccelli per lo più si dimostrano nei moti e nell'aspetto lietissimi; e non da altro procede quella virtù che hanno di rallegrarci colla vista, se non che le loro forme e i loro atti, universalmente, sono tali, che per natura dinotano abilità e disposizione speciale a provare godimento e gioia: la quale apparenza non è da riputare vana e ingannevole." (Leopardi 1895, pp. 273–274).
76. "Veramente molto conforto e diletto ci porge, e non meno, per mio parere, agli altri animali che agli uomini, l'udire il canto degli uccelli. E ciò credo io che nasca principalmente, non dalla soavità de' suoni, quanta che ella si sia, né dalla loro varietà, né dalla convenienza scambievole; ma da quella significazione di allegrezza che è contenuta per natura, sì nel canto in genere, e sì nel canto degli uccelli in ispecie. Il quale è, come a dire, un riso, che l'uccello fa quando egli si sente star bene e piacevolmente." (Leopardi 1895, p. 276). Cf. Marzot 1966, on Leopardi's philosophy of laughter and the transformative powers of its articulation in his thoughts on nature, human vs. non-human existence, and literature.
77. "Onde si potrebbe dire in qualche modo, che gli uccelli partecipano del privilegio che ha l'uomo di ridere: il quale non hanno gli altri animali; e perciò pensarono alcuni che siccome l'uomo è definito per animale intellettivo o razionale, potesse non meno sufficientemente essere definito per animale risibile; parendo loro che il riso non fosse meno proprio e particolare all'uomo, che la ragione." (Leopardi 1895, pp. 276–277).
78. Buffon 1820, p. 67.
79. "Avendo l'udito acutissimo, e la vista efficace e perfetta in modo, che l'animo nostro a fatica se ne può fare una immagine proporzionata; per la qual potenza godono tutto giorno immensi spettacoli e variatissimi, e dall'alto scuoprono, a un tempo solo, tanto spazio di terra, e distintamente scorgono tanti paesi coll'occhio, quanti, pur colla mente, appena si possono comprendere dall'uomo in un tratto; s'inferisce che debbono avere una grandissima forza e vivacità, e un grandissimo uso d'immaginativa. Non di quella immaginativa profonda, fervida e tempestosa, come ebbero Dante, il Tasso; la quale è funestissima dote, e principio di sollecitudini e angosce gravissime e perpetue; ma di quella ricca, varia, leggera, instabile e fanciullesca; la quale si è larghissima fonte di pensieri ameni e lieti, di errori dolci, di vari diletti e conforti; e il maggiore e più fruttuoso dono di cui la natura sia cortese ad anime vive" (Leopardi 1895, pp. 281–282).
80. The text of Plotinus (with excellent translation into Italian by Roberto Radice) on which I have based my analysis is Plotinus 2002, a magnificent, unique modern edition which conveniently presents the whole *Enneads* in one annotated volume with facing Greek.
81. Leopardi 1987–1988, II, pp. 76 ff.
82. Leopardi 1987–1988, I, p. 164.
83. In the poem "Alla primavera, o delle favole antiche" [To the Spring, or the ancient fables]. Leopardi 1987–1988, I, pp. 33–35.
84. The celebrated verse in Italian is as follows: "…atra face del ver". From the poem "Alla primavera, o delle favole antiche". Leopardi 1987–1988, I, p. 33.
85. Selcer 2010.
86. Eliade 1961, p. 49.
87. Cioran 1987, p. 10.

88. Cioran 1987, p. 10.
89. "Nos vero inter res caste et perpetuo versantes, intellectum longius a rebus non abstrahimus quam ut rerum imagines et radii (ut in sensu fit) coire possint; unde fit, ut ingenii viribus et excellentiae non multum relinquatur. Atque quam in inveniendo adhibemus humilitatem, eandem et in docendo sequuti sumus." (Accessible online at: http://www.thelatinlibrary.com/bacon/bacon.praefatio.shtml).

Bibliography

Bernabò Brea, Luigi. 1998. *Le maschere ellenistiche della tragedia Greca.* Naples: Publications du Centre Jean Bérard. Retrievable online at: http://books.openedition.org/pcjb/266.

Bossuet, Jacques Bénigne. 1863. *De la connaissance de Dieu et de soi-même. Traité du libre arbitre—Logique—Traité des causes.* Paris : Librairies de L. Hachette.

———. 1875. *De la connaissance de Dieu et de soi-même.* Nouvelle Édition. Paris: Librairie classique d'Eugène Belin.

Boulogne, Jacques. [Ed.] 2005. *Le Grecs de l'antiquité et les animaux. Le cas remarquable de Plutarque.* Lille: Collection travaux et recherches, Université Charles-De-Gaulle Lille 3.

Buffon, Georges-Louis Leclerc, Comte de. 1820. *Le opera di Buffon nuovamente ordinate ed arricchite della sua vita e di un ragguaglio dei progressi della storia naturale dal MDCCL in poi. Prima edizione Italiana adorna di nuove e diligenti edizioni.* Volume 17. Venice: Al negozio di libri all'Apollo.

Cicero. 1552. *De legibus libri III. In eosdem commentarii Adr. Turnebo auctore.* Paris: Ex Officina Adr. Turnebi.

———. 1853. *The treatises of M. T. Cicero: On the Nature of the Gods; On Divination; On Fate; On the Republic; On the Laws; and On Standing for the Consulship.* Translated and edited by C. D. Yonge. London: Henry G. Bohn.

———. 1872. *De legibus libri tres.* Recongnovit Reinholdus Lotz. Leipzig: B. G. Teubner.

Cioran, Emil M. 1987. *Aveux et anathèmes.* Paris: Gallimard.

Corti, Alfonso. 1851. "Recherches sur l'Organe de l'Ouïe". *Zeitschrift für Wissenschaftliche Zoologie* 3, pp. 109–169.

Delamarre, L., and Brunetière, F. 1907. "Jacques-Benigne Bossuet". In *The Catholic Encyclopedia.* New York: Robert Appleton Company. Retrievable online at: New Advent: http://www.newadvent.org/cathen/02698b.htm.

Della Porta, Giovan Battista. 1586. *De humana physiognomonia libri IIII.* Vici Aequensis: Apud Iosephum Cacchium.

———. 2013. *Della fisionomia dell'uomo libri sei.* Edited by Alfonso Paolella. Volume II. Edizione Nazionale delle Opere di Giovan Battista della Porta. Naples: Edizioni Scientifiche Italiane.

Derrida, Jacques. 2008. The animal that therefore I am. Edited by Marie-Louise Mallet. Translated by David Wills. New York: Fordham University Press. (Original edition in French: *L'Animal que donc je suis.* Paris: Éditions Galilée, 2006).

Eliade, Mircea. 1961. *Images and symbols: Studies in religious symbolism*. Princeton: Princeton University Press [or. ed. *Images et symbols*, Paris: Gallimard, 1952].

———. 1964. *Shamanism: archaic techniques of ecstasy*. London: Routledge & K. Paul.

Flourens, Pierre. 1842. *Recherches expérimentales sur les propriétés et les fonctions du système nerveux dans les animaux vertébrés*. Paris: Chez J. B. Baillière.

Foucault, Michel. 1999. "The meaning and evolution of the word parrhesia". In *Discourse and Truth: the problematization of parrhesia*. Digital Archive: Foucault.info. Available online at: https://foucault.info/doc/documents/parrhesia/foucault-dt1-wordparrhesia-en-html.

Gabrieli, Giuseppe. 1932. "Bibliografia Lincea. Giambattista Della Porta. Notizia bibliografica dei suoi mss., e libri, edizioni, ecc., con documenti inediti". *Rendiconti dell'Accademia Nazionale dei Lincei, Classe di Scienze Morali, Storiche e Filologiche*, series 6, volume 8, pp. 206–277.

Galilei, Galileo. 1623. *Il Saggiatore*. Rome: Giacomo Mascardi.

———. 1880. *The sidereal messenger of Galileo Galilei and a part of the preface to Kepler's Dioptrics containing the original account of Galileo's astronomical discoveries*. Translated with Introduction and Notes by Edward Stafford Carlos. London: Rivingtons.

Gleich, O., and Langemann, U. 2011. "Auditory capabilities of birds in relation to the structural diversity of the basilar papilla". *Hearing Research* 273, pp. 80–88.

Haar, Alfred. 1910. "Zur Theorie der orthogonalen Funktionensysteme. (Erste Mitteilung)". *Mathematische Annalen* 69, pp. 331–371.

Helmholtz, Hermann von. 1863. *Die Lehre von den Tonempfindungen, als physiologische Grundlage für die Theorie der Musik*. Braunschweig: F. Vieweg und Sohn.

———. 1877. *Die Lehre von den Tonempfindungen, als physiologische Grundlage für die Theorie der Musik*. Vierte Aufgabe. Braunschweig: F. Vieweg und Sohn.

———. 1885. *On the sensations of tone as a physiological basis for the theory of music*. Translated by A. J. Ellis. 2nd English ed. London: Longmans & Co.

Jablonca, Eva, and Lamb, Marion J. 2005. *Evolution in four dimensions. Genetic, epigenetic, behavioral, and symbolic variation in the history of life*. Cambridge, MA, London, England: The MIT Press.

Köppl, Christine. 2011. "Birds e same thing, but different? Convergent evolution in the avian and mammalian auditory systems provides informative comparative models". *Hearing Research* 273, pp. 65–71.

Legée, Georgette. 1992. *Pierre Flourens, physiologiste et historien des sciences*. 2 vols. Abbeville: F. Paillart.

Leopardi, Giacomo. 1895. *Operette morali*. Edited by Nicola Zingarelli. Naples: Luigi Pierro.

———. 1987–1988. *Poesie e prose*. 2 vols. Milan: Mondadori.

Levi, Primo. 1991 [1986]. *I sommersi e i salvati*. Turin: Einaudi.

Levinas, Emmanuel. 1957. "La philosophie et l'idée de l'Infini". *Revue de Métaphysique et de Morale*, 62e Année, No. 3, pp. 241–253.

———. 1982. "La souffrance inutile". *Giornale di metafisica* 4, pp. 13–26.

———. 1991. *Entre nous. Essais sur le penser-à-l'autre*. Paris: Bernard Grasset.

———. 1998a. *Entre nous. Thinking-of-the-other*. Translated by Michael B. Smith and Barbara Harshav. New York: Columbia University Press.

———. 1998b. *Tra noi. Saggi sul pensare-all'altro.* Translated by Emilio Baccarini. Milan: Editoriale Jaca Book.

Marzot, Giulio. 1966. *Storia del riso Leopardiano.* Messina and Florence: G. D'Anna.

Nardi, Bruno. 1967. *Saggi di filosofia Dantesca.* Second Edition. Florence: La Nuova Italia. [1st ed. 1930, Milan: Società anonima editrice Dante Alighieri].

Nettl, Bruno. 1974. "Thoughts on improvisation: A comparative approach". *The musical quarterly* 60, pp. 1–19.

Newmyer, Stephen T. 2006. *Animals, rights, and reason in Plutarch and modern ethics.* New York and London: Routledge.

Palmieri, Paolo. 2016. *Hermes and the telescope. In the crucible of Galileo's life-world.* New York: Peter Lang.

Pascoli, Giovanni. 1902. *La mirabile visione. Abbozzo d'una storia della Divina Comedia.* Messina: Vincenzo Muglia.

Pianigiani, Ottorino. 1907. *Vocabolario etimologico della lingua italiana.* Rome and Milan: Società editrice Dante Alighieri.

Plotinus. 2002. *Enneadi.* Translated by Roberto Radice with facing Greek text. Introductory essay, prefaces and annotations by Giovanni Reale. Including the *Life of Plotinus* by Porphyry, edited by Giuseppe Girgenti. Milan: Mondadori.

Plutarch. 1957. *Moralia. Volume XII.* Translated by Harold Cherniss and William Helmbold. Cambridge, Massachusetts: Harvard University Press.

Sawyer, R. Keith. 1992. "Improvisational creativity: An analysis of jazz performance". *Creativity Research Journal* 5, pp. 253–263.

———. 1996. "The semiotics of improvisation: The pragmatics of musical and verbal performance". *Semiotica* 108, pp. 269–306.

Selcer, Daniel. 2010. *Philosophy and the book: Early modern figures of material inscription.* London and New York: Continuum.

Turner, Steven R. 1977. "The Ohm-Seebeck dispute, Hermann von Helmholtz, and the origins of physiological acoustics". *The British Journal for the History of Science* 10, pp. 1–24.

On Birdsong: An Essay on Animal Physiology and Biology in Relation to Sexual Selection and the Struggle for Existence by Luigi Paolucci

EDITED AND TRANSLATED BY PAOLO PALMIERI

PART ONE

Comparative Phenomenology of Birdsong

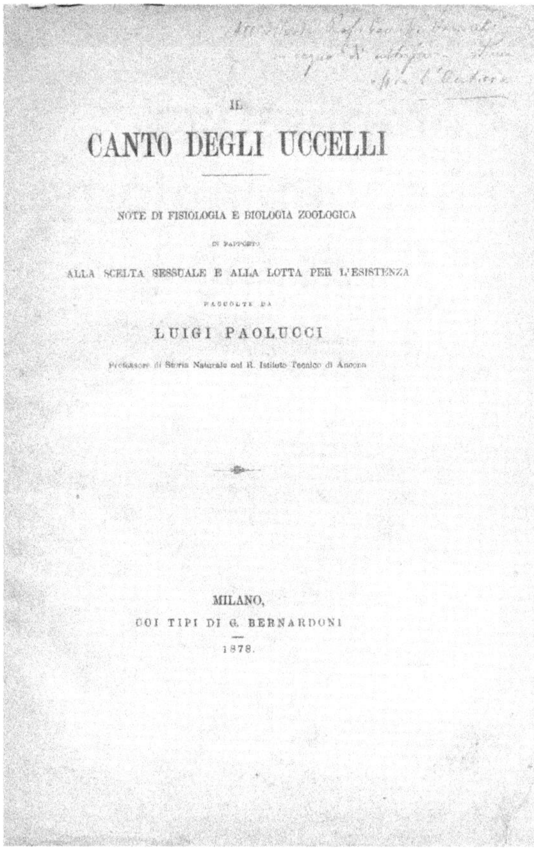

FIGURE 1.1 The frontispiece of Paolucci's monograph on birdsong. From a copy in my personal collection. On the top right hand side of the page a manuscript inscription appears in the hand of Paolucci. It is dedicated to "V. Cerruti in segno di altissima stima…" [to V. Cerruti as a token of the highest esteem…]. In all likelihood, the dedicatee was the engineer and mathematician Valentino Cerruti. Figure and caption by Paolo Palmieri.

PART ONE, CHAPTER ONE

On Natural Language

By the word *language* it is commonly understood the faculty, with which only humans are endowed, of communicating ideas by means of certain articulated sounds referred to as 'words'. However, it would have been useless to accept this definition in my work, since I ultimately intended to research what it is in birds that is functionally equivalent to human language. Thus, I adopted and consistently made use of the word *language* in order to indicate any means of communication among beings that need to form cognitive relationships.

It is true that some people think that insofar as humans represent the supreme vertex of the great zoological tree, owing both to their bodily organization and to their brain functions, they do not at all belong in the animal kingdom, precisely because humans alone possess what we call *speech*, *word*, *language*. For this reason humans would form a separate kingdom, and the science concerning them would be an independent discipline called 'anthropology'. Doubtlessly human speech is a marvelous power, by means of which our relationships become concrete, actual, and are brought to perfection. It is nowadays the necessary instrument for communicating ideas. By long habit it has become the auxiliary yet indispensable instrument for cognition. The more perfect its organization the more civil the human culture that speaks it. Yet students of human language have gone astray in speculating about its origins as they contemplated its sublime purposes, admiring its economy and praising its effects, but had scruples about analyzing its elements, or I should rather say its organs, as a Muslim would have about anatomizing a human cadaver. For, language is an organic entity, well defined and possessing an evolutionary life, the product of the most perfect creature. Hence, since in the science of language the rules of etymology were misconceived at the very beginning and the second causes of language were poorly understood, if not wholly misunderstood, the first cause of language was assumed to be that to which mankind had had recourse on many other occasions, that is, it was accepted that language had a divine origin (what Johann Süssmilch called *the gift of the word*).[1]

Others, equally arrogating to themselves the power to pass a verdict on these issues, claimed that speech was produced by mankind, as a discipline invented consciously by mutual agreement.[2] Accordingly, they laid out its norms a priori, forming a doctrine that is founded on conventionalism.

These theories on the origin of speech, which I would call *literary deductions*, held sway, especially among the learned, until the science of language, a newcomer that has so far failed to find the right word to define itself, entered the vast field of natural science, to which it is very closely related in virtue of countless analogies. Indeed, Max Müller would like to categorize it as a natural science. Initially, Wilhelm von Humboldt, Franz Bopp, and Jacob Grimm, and later on Max Müller, William Whitney, Graziadio Isaia Ascoli laid the foundations of the study of language in keeping with the same methodology by which George Cuvier, Richard Owen, Alexander von Humboldt and Charles Lyell had created modern comparative anatomy and geology. Thus, by following in the footsteps of ancient hypotheses on the genesis of human language (such as Epicurus's), scholars came to investigate its origin by means of correct analyses, which in contrast to the former ones I would call *scientific deductions*. As the functions were discovered from the exact knowledge of the forms, language was caught, so to speak, in its vital moments, and it became clear that it lives in a process of continuous evolution and incessant metamorphosis, so that no sooner does it become worn out than it rejuvenates again. By the same token, organized beings live, decay and rejuvenate again, in precisely the same fashion as the species is transformed. As animal life would in the last analysis consist of the incessant oxidation of its tissues (according to Wilhelm Kühne), so language would thrive in the process of phonetic corruption and dialectal regeneration (according to Max Müller).

Once the science of language, or phonology, or glottology, as one prefers to call it, entered the field of natural science, scholars quickly realized that one could attribute to the well-defined elements of speech, namely, words, a similar origin to that which most zoologists and botanists attribute to animals and plants. That is, a natural origin.

Human beings felt a need to relate to other human beings as soon as they came into the world, as they were thrown into the struggle for existence by nature, a struggle that would quite naturally have been terrifying for them initially, as they were among the most unwarlike creatures. Everything that acts upon each one of the following senses, hearing, sight, and touch, answers the purposes of the cognitive relation among humans and between humans and animals. Indeed, in addition to hearing, the means by which we hear the sounds that compound words, the sense of sight makes us perceive the so-called *expressive signs*, by which humans, both civilized and uncivilized (according to John Lubbock), give more or less force to speech. Sight is the only means by which deaf people communicate with us, and we understand their peculiar language, which can only be mimetic unless a special education is imparted to those unhappy people. Finally, among the human means of communication, we find touch although it is much more common in animals than humans. Thus, since sight, hearing, and touch are efficacious in making us understand those acts that signify an idea, or a portion of it, which are expressed with voice, or with signs, or with contact, humans might have forged a semiotic language, or a tactile one, precisely as they constructed a phonetic one. Humans created a phonetic language by free choice as soon as they realized that this type was the readiest, most convenient, and the best (as suggested by Charles Darwin, and Giuseppe Bagarotti).[3] They abandoned the other options to which, however, they must have recourse when they cannot hear words, as in the case of the deaf, or when they can neither hear nor speak as in the case of people who are both hearing and speaking impaired.

Let us remember that certain natural patterns that are common to humans and which depend on will (such as curiosity, attention, and imitation), or on special circumstances of the nervous centers (such as anger, pain, and cheerfulness), are also manifested by other animals, not to mention the fact that there are intellectual faculties that are shared by both humans and animals. As in humans these patterns are expressed by means of the voice, or more rarely by body language and touch, so it seems to me that the same phenomenon occurs in animals. But of course this happens only in those animals that do not descend below a certain level of degradation in the hierarchy of organisms, for when the capacity for creating the simplest patterns of cognitive behavior is missing, then the expression of those patterns will obviously not take place.

Assuming that there are three forms of language in humans, namely, phonetic, semiotic, and tactile, I will attempt to demonstrate, with the help of facts selected from zoology, how animals may take advantage of the same expressive means in social life.

1. Tactile Language

If in certain animals that have a low level of organization (such as *Coelenterates*, echinoderms, cystoids,[4] and *Rhizopoda*) nature had produced an aberration by excluding these animal forms from the laws of evolution and especially from developmental laws, instead endowing them with a nervous center (brain) but at the same time preserving the rudimentary organs with their imperfections, I think that we would have seen the ambulacral tube-feet of the *Spatangus*, the cirrous arms of the ophiuroids, the tentacles of many ctenophores, *Acalephae*, hydra and polyps, the filiform appendage of the *Noctilucales*, the pseudopods of the *Foraminifera*, etc., all serving to express and intend by means of touch that which other animals manifest and conceive of by means of signs through sight and by means of the voice through hearing. In such animals at low levels of organization this phenomenon would have occurred necessarily, by an eliminative process, as the organs of phonation, of hearing, and of sight, which are more fitted to the expression and perception of brain functions, are missing. However, lower animals are not capable of these higher functions. On the other hand, those which are endowed with higher faculties prefer the most comfortable means. This is the reason why, I believe, animals which somehow communicate among themselves through the use of touch are rare. But it is also the case that only the simplest ideas can be expressed by touch, while cognition becomes more complex and specialized in the presence of hearing and sight.

Here are a few examples of tactile language taken from the natural history of animals. Pierre Huber devotes many pages to the language of ants, which he thinks occurs really in virtue of the tentacles by means of which two ants, touching each other in countless ways, recognize each other, alert each other, help each other, or throw down the gauntlet of a mortal duel.[5] Sir John Lubbock, quoted by Prof. Darwin, in describing the male of the *Smynthurus luteus* (*Thysanura*) recounts that

> it is very amusing to see these little creatures coquetting together. The male, which is much smaller than the female, runs round her, and they butt one another, standing face to face, and moving backward and forward like two playful lambs. Then the female pretends to run away and the male runs after her with a queer appearance of anger, gets in front and stands facing her again; then she turns coyly round, but he, quicker and more active, scuttles round too, and seems to whip her with his antennae; then for a bit they stand face to face, play with their antennae, and seem to be all in all to one another.[6]

Butterflies too might furnish many examples of cognitive relations made possible by touch, both between male and female and among individuals of the same sex.

These are mute animals and for them the natural choice for sexual selection was the splendid coloring of body and wings. I was able only to observe a few species. In the last form of the silkworm (*Bombix mori*), the male goes round the female repeatedly before winning her over, while he seems to prepare her for the conjugal act by touching her with a delicate and frisk fluttering of the wings. The mating of our common six-spot burnet (*Zygaena filipendulae*) is preceded by a reciprocal touching of the heads with the antennae. Among small white butterflies (*Pieris rapae*), painted ladies (*Vanessa cardui*), swallowtail butterflies (*Papilio podalirius, Papilio machaon*), the common clouded yellow butterflies (*Colias edusa*), and possibly in many other species, during the casual encounter, or in duels, or when congregating together, I often saw one of the combatants or lovers use their first pair of anterior legs in any manner of way, as if fondling, or displaying benevolence, or belligerence. But these examples selected from lower animals, in which the power and need to communicate their minimal, instinctive sentiments among themselves are almost nil, might confuse us. We risk mistaking the merest excitation of the senses through touch, occurring frequently among many animals and always serving a pre-determined purpose in a fixed schema of activity and passivity, for real tactile language. Yet this doubt will be alleviated as we consider the higher levels of the zoological scale.

As for fish and many other aquatic animals, about whose biology we will learn enough only from studying them in aquariums, we know nothing that might suggest that they make use of a tactile language. Yet there are myriad animals, thousands of species, often endowed with a high degree of organization, which are incessantly exercising all the functions that help both the preservation of the individual and the permanence of the species. Might nature not have provided those mute dwellers of the waters with a tactile faculty suitable for communication, given that they must always defend themselves from predators, or scavenge for food when impelled by hunger, or court the females, carried away by the instinct for love, and undertake long voyages as they are governed by laws of localization whose present patterns are still little understood?[7] Many accessory organs of fish that Darwin and his followers would consider as indirect agents of the evolution of species, or useful adornments for sexual competition, might also be used as instruments for touch among individuals. A few examples will suffice. Red mullets (*Mullus barbatus*), striped red mullets (*Mullus surmuletus*), and the *Upeneus* of the Indian seas (*Upeneus trifasciatus*) carry mobile cirri in the corner of the inferior jaw. The St. Peter's Fish (*Zeus Faber*) has the middle-spinal lobes of the dorsal fin lengthened in delicate, wavy ribbons. The common angler (*Lophius piscatorius*) has a true tentacle stemming from the nasal region. The cornet fish (*Fistularia tabacaria*) carries a filiform extension, beyond the caudal fin, whose length is about one-half the length of the body. The red lionfish of the Indian Ocean (*Pterois volitans*) is adorned with three pairs of tentacles on the superior lip, on the nose and the forehead, in addition to the numerous ramifications of the thoracic and dorsal fins. Two types of saw-like horns rise above the forehead of the butterfly blenny (*Blennius ocellaris*). Among the sheatfish, the Egyptian *Sciarmut*[8] (*Heterobranchus anguillaris*), the electric catfish (*Malapterurus electricus*), known among us as the *Raasch*[9] of the Arabs and Senegalese, have the external perimeter of the lips rich in worm-like appendages. I must reiterate, though, that there is no conclusive evidence suggesting that these organs could be employed as means of tactile communication.

Among the reptiles, the mobile and vibratile tongue of the ophidians, as well as the extensible one of certain saurians, must be regarded as tactile organs of great importance when one considers the horn-like armor that covers these animals. These organs of touch might well stand in for the most common means of communication (sounds) of which these animals are deprived almost entirely.

All the examples mentioned so far lead us to speculate that, where vocal and semiotic means of communication were lacking (such as in less perfect animals), touch was employed as an agent mediating cognitive activities that can be carried out also by rudimentary organisms. Humans and other higher-order animals, which are capable of using voice and gesture in order to communicate, have abandoned tactile language since the beginning of their existence. It was the lack of use that in this case eliminated the special meaning of the function of an organ (in Darwin's view). However, only humans have created speech. And as they have exploited all possible means to share their ideas, they also preserved vestiges of tactile language whose effectiveness has in the progress of civilization been perfected to the point of expressing the most subtle thoughts and the deepest feelings from the heart. How much is communicated by the speechless shaking of hands with a true friend, or an honest patron, or a magnanimous benefactor? How much is said by the anxiety of the desperate mother, who has no words to express her overwhelming affection, and yet is so eloquent in touching the face, the forehead, and the neck of the beloved child, hugging him to her bosom? How much is told by the first kiss that uncovers the sweet mystery of love to a girl, and which sometimes condenses a longest story? These considerations might sound preposterous. Yet, in the final analysis, they appear to be true if we make the assumption that all sorts of action carried out on the peripheral apparatus of the sensory nervous system might become the sign of an idea, once the action has been established as being conventional, or made significant by protracted usage. I will now turn to analyzing semiotic language briefly.

2. Semiotic Language

Semiotic language, based on signs or mimicry, requires a high degree of organization for it to be both conceived of and crafted, as well as for it to be understood. Its importance is greater than that of a rudimentary touch-language, in fact it is almost as important as speech and phonetic language, especially when it has been perfected by sophisticated human cultures. Indeed, considering the cognitive apparatus that often must preside over the production and comprehension of semiotic language, I would not hesitate to regard it as superior to speech, at least in certain cases. In fact, there is no such thing as a semiotic language among the inferior and the lowest animals, and it is only vaguely recognizable in superior animals.[10]

Humans took advantage of semiotic language. But perhaps it is puzzling that, when we consider animals that are closest to humans in the scale of vertebrates, namely, mammals, we find that their natural talent for semiotic language is rather modest, while it is fully recognizable in birds, which are gentle creature, eminently esthetic, though perhaps only schematically realized in their behavior. The reason for this discrepancy must be found in the principal purposes that every language serves, including semiotic language. These purposes, according to the theories of Alfred Russel Wallace, Charles Darwin, Ernst Haeckel, and Giovanni Canestrini, would be two, namely, the struggle for existence and sexual competition.[11] In the case of birds, we observe the very peculiar shaking of the head of some nocturnal predatory species such as the owl (*Strix passerina*) and the barn-owl (*Strix flammea*). It cannot be denied that this sign is perceived also by other birds, as we all know that, when the owl shows up, we hear the call of motley sparrows crowding around this ludicrous impersonator to mock its fawning gestures and frantically flying around in a frenzy of laughter. In the oscines (i.e., songbirds) the obsequious bows are well known in which the robins (*Sylvia rubicola*) indulge when they alight on a branch or a turf. I observed that these manifestations of reverential behavior form rhythmic patterns especially before the fighting between two individuals, typically occurring in winter, for dominion over a hedge, or a garden, or brushwood. The two combatants take position close to each other, they flutter repeatedly in the manner I have just described, then after vocalizing a special hiss they go after each other doggedly. However, it remains controversial whether these quick motions of the throats are to be considered rather a means of communication than, for instance, a sheer display of beauty, as is observed in the peacock (*Pavo cristatus*) when it fans out the tail, or in the scarlet macaw (*Ara macao*) when it balances on a branch and spreads the beautiful wings and the tail, or in the hoopoe (*Hupupa epops*) when it waves the sincipital crown. Indeed, the robin summarizes in its red throat the entire humble beauty of its dress. Likewise the blue-throat robin (*Sylvia seucica*), the stonechats (S*axicola oenanthe*), the European stonechats (*Saxicola rubicola*), and the common rock thrush (*Monticola saxatilis*) lavish extravagant salutations even though their reddish or bluish shirts hardly make them particularly charming. Other birds such as the wagtails (*Motacillae*), the redstart (*Ruticilla phoenicura*), the long-tailed bushtit (*Parus caudatus*) often shake their tail downwards, while walking or resting. Others such as the common blackbird (*Turdus merula*), ring ouzel (*Turdus torquatus*), the nightingale (*Luscinia philomela*), and the common wren (*Troglodytes Europaeus*) raise the tail almost vertically. Many finches move it laterally. The northern mockingbird (*Mimus polyglottos*) widens and closes the tail while both hopping around and flying. All these examples might also be interpreted as suggesting that the motions of the tail in birds, before becoming instinctive by inheritance, might have served to put to flight the insects hiding in the soil and among leaves, so as to catch them more easily.

Empirical studies of the signs used by animals as means of communication are scarce and not always reliable. Semiotic language became an art made possible only by civilization for the benefit of human beings affected by the misfortune of deaf-mutism. I would not claim that in such a language there is absolutely nothing natural. But the fact that it has not been found in animals, or only traces of it, might confirm it. On the other hand, a cursory examination of the common elements of semiotic language found among almost all cultures in the business of daily life, would show that this art is conventional, and its origin must have been an arbitrary act of free will.

'No' is expressed by the Italians and many other Europeans with a motion of the head turning around the vertical axis. A Neapolitan says 'no' with a slight pursing of the lips rising upwards, while a Turk does it in the same way as we say 'yes'. Oriental people place their hand on the heart instead of shaking it as when greeting friends. A Chinese respectfully puts his hat on. Polynesians, Malaysians, and other Asian people feel the obligation to sit before a high-ranking person even though they are cultured races.[12] For a native of Vatavulu,[13] of Congo,[14] among the Wahuma of the East,[15] and in central Africa,[16] it would be highly inappropriate not to turn one's back towards a distinctive individual to whom one speaks. The inhabitants of Mallicolo (according to James Cook's account) express their admiration for a man by hissing at him.[17] Among the Toda (according to Lubbock) one pays one's respect to somebody by waving the hands in same manner in which our children mock one another. Examples abound.

FIGURE 1.2 A disturbing yet fascinating engraving, a picture from our collective unconscious, or perhaps a censored reminiscence refusing to condone the savagery of cultural and political imperialism? It portrays a native man from Mallicolo. It was published in Cook's book. The man is dressed up in luxurious gyrations of drapery that make him look like an imposing figure in a classical-master study, showing little of his ape-likeness and ugly nudity. For, according to Cook, "they are the most ugly, ill-proportioned people I ever saw...The men go quite naked, except a piece of cloth or leaf used as a wrapper" (Cook 1777, II, p. 34). Indeed, as Aldous Huxley put it, "[c]ivilized human beings wear clothes, therefore there can be no portraiture, no mythological or historical story-telling without representation of folded textiles" (Huxley 2009 [1954], p. 30). And perhaps there can be no story-telling in European science without the racist portrayal of the 'other', the 'uncivilized' savage, consciously or unconsciously disguised in the foldings of abstraction and logical argument. I could not resist the inexplicable urge to bring to bear on the reader's life-world this picture-wonder, a found object staring at my consciousness, a discard on the scrapheap of my research in the history and philosophy of science and a bruise from my philological struggles, that might open up unsuspected possibilities of transformation and understanding, as it did for me, for in the words of the poetess Edith Doron "world, as that which is hidden in plain view, must be torn asunder, brought to the fore in order to free the being of that silent object which has 'agreed to keep quiet about us' from an imperial interpretive statement. If we are to understand encounter not as an act of comprehension in the sense of 'grasping the world' but instead as the immersive experience of de-familiarization, a dislodging of the concealed given ground of world, then the possibility of transformation is freed" (Doron 2008, pp. iii–iv). Figure and commentary by Paolo Palmieri. Picture from: *The Graduate Institute, Geneva—The Kathryn and Shelby Cullom Davis Library. Remarkable Items—Diplomacy—Rare Books and Items (17th-20th c.).* Permalink: http://digitalcollections.graduateinstitute.ch/idurl/1/1333 (Figure and caption by Paolo Palmieri).

From the above discussion we can draw the conclusion that semiotic language is at most rudimentary in animals, and in general that when we find it in humans it is artificial.

3. Phonetic Language

The means that both animals and humans employ with the greatest success in forging cognitive cultures is neither tactile not semiotic language, rather it consists in uttering sounds, words, harmonies, shouts, and noise, even though, as we have tried to show, they occasionally have recourse to touch and to special bodily movements. Communicative sounds display the most bewildering variety of effects, ranging from the divine persuasiveness of human eloquence, through the mollifying captivation with the melodies played by a Paganini, to the terror caused by a roaring lion, or the hatred for the stupefying, relentless clamor of cicadas.

The organs that animals and humans use to produce sound are most diverse, often elastic, tensioned cords, or pipes, or vibrating cavities, sometimes tender expansions like vesicles, but also stiff reeds, indented bows, drum-membranes, and thin laminas. Yet the physical mechanism of sound-production is always the same, namely, the vibration of an air column more or less enclosed in a limited volume. Also the means of perception is the same, namely, the acoustic nerve. Hence in the inferior and least organisms that have no auditory apparatus at all, or only an embryonic one, sounds would have been useless insofar as they serve cognitive communication. Indeed, the burgeoning life of the oceans is mute. Their most perfected inhabitants, fish and mollusks, are often endowed with a vesicle containing otoliths, which anatomists describe as an 'ear', but which the physiologists regard as a poor semblance of an auditory organ. It remains controversial to what extent fish and mollusks can hear. Their auditory organ barely perceives noises transmitted in the air. On the other hand, their element is not the most suitable for the propagation of acoustic waves. Thus, the language of sound is the sole preserve of animals breathing air. It is common in humans, mammals (except the cetaceans that lack the vocal cords), in birds, reptiles, in many batrachians and insects. Although the objective of sound production in animals appears to me to be always the same, the difference in degree is immense. The reason is the exceptional variety in cognitive styles. In this regard the most savage human being is far removed from the most cultured yet non-human mammal.

Yet in regard to the simplest cognitive abilities in humans, which are the most complex in animals, phonetic language might be considered as equivalent in both. In the remainder of the book, I will try to demonstrate that this is the case precisely in birds. Let the following be clear, at least. Although one might argue that animals do not speak, it can hardly be denied, except perhaps by a mind numbed by prejudice, that a lamb will desperately call its mother before being slaughtered.

Before proceeding a fundamental division of the signs used as phonetic language by humans and animals must be made on the basis of the structure of the vocal organs. In warm-blooded vertebrates the vocal organs consist of an apparatus known as 'larynx', situated in the superior extremity of the trachea, or in the base, which is in direct communication with the lungs or other air cavities, and which is capable of contraction and expansion at will, thus allowing speaking, singing, and shouting. In all other animals, these contraptions of sound-based language are diversely located throughout the body. As we shall see in more detail in the next chapter, they look like musical instruments, little trumpets, violins, or organ pipes in miniature. Having said this we draw the conclusion that the sounds produced in the larynx are modifiable in timbre, pitch, and intensity, at the will of the animal, within a more or less extended range. On the other hand, sounds that can only be made by special mechanisms having either fixed parts or parts that are hardwired, so to speak, cannot be modulated by the animal, and therefore such sounds would resemble those produced by a key of the organ, or a given position in the violin. Thus, among the animals that can produce sounds some have the predisposition to modify them by transforming and separating them as they wish, others appear to be captivated by the same song or by slight variations all their life. In the first group we find humans, mammals, and birds, in the second all other animals capable of articulating a phonetic language. Since we have not been preceded by anybody else in these considerations, we will coin two new words to distinguish these two groups, so that we will partition the animal kingdom in regard to the nature and predisposition of their voice. We suggest to call the first group 'heterophonous' (by conjoining the Greek words eteros = *other* and fone = *voice*), that is, endowed with mutable voice, and the second 'autophonous' (autos= *same* and fone = *voice*), that is, with immutable voice.[18]

Notes

1. {Cf. Süßmilch 1776, to which Paolucci is probably referring, possibly indirectly as he does not reference any essay by Süßmilch in detail. The original title of Süßmilch's work, which Paolucci may have read, is *Versuch eines Beweises, daß die erste Sprache ihren Ursprung nicht von Menschen, sondern allein vom Schöpfer erhalten habe* (Attempt at a demonstration that the first language was created not by man but by the Creator). According to Avi Lifschitz, "[d]espite the theological connotations of the title, this [i.e., Süssmilch's] essay was a shrewd attempt to present a philosophical proof 'from the realm of nature' against the

naturalistic-Epicurean theory of the emergence of language" (Lifschitz 2012, p. 83). Note that in Paolucci's text *Süßmilch* is spelled *Süssmilch*.}

2. {I have rendered the original *arte* with *discipline*, as in Paolucci's work the word 'arte' seems to convey the sense, more common in nineteenth-century Italian than nowadays, of a complex set of rules established in order to accomplish something in an orderly fashion, which suggests that *arte* means not only the body of theoretical definitions of the norms but also the skills needed to apply them competently.}
3. {It is worth mentioning Giuseppe Bagarotti's little-known but interesting philosophical essay on the human intellect (Bagarotti 1857), which is most likely the work that Paolucci has in mind here. In that essay, Bagarotti discussed language, the relationship between thought and language, and the nature of signs. He proposed a general linguistic theory according to which language includes speech, pantomime, telegraphy, military codes, hieroglyphics, emblems, Chinese and Asian pictograms, and algebraic symbolism. See Bagarotti 1857, pp. 97 ff.}
4. {I assume that some cystoids might belong in the species that Paolucci refers to as 'mixocestoidi', for which I could not find an exact English cognate.}
5. Huber 1810.
6. Darwin 1871b, p. 252. {English text from Darwin 1871a, I, p. 348.}
7. {Cf. Stoppani 1873, p. 143. Antonio Stoppani, the founding father of modern Italian geology, had thoroughly discussed biogeography, what he called the phenomenon of the 'accantonamento' (localization) of the world faunas. Paolucci made use of the same word 'accantonamento', referring to the ideas that Stoppani had popularized in his monumental treatise on geology, which was obviously well known to Paolucci, namely, the *Corso di geologia* in three volumes. See Zanoni 2014, for an excellent introduction to the life and work of Stoppani, and his influence on Italian cultural life, politics, religion, and science. On the geographical distribution of species, it is possible but I think unlikely that Paolucci might have consulted Wallace 1876.}
8. {A transliterated noun used by Paolucci with no equivalent in English.}
9. {A transliterated noun used by Paolucci with no equivalent in English.}
10. In the zoological literature it became customary to make use of the phrases 'inferior, and lowest animals' and 'superior animals' in order to categorize animals according to the greater or smaller degree of differentiation of their organs and functions. However, this artificial division does not seem to me to have been perfectly clarified, let alone uniformly understood by everybody. Thus, while using the aforementioned terminology, I wanted to include in the superior animals all the vertebrates, or hypocotyledons of Van Beneden (Van Beneden 1853), in the inferior, mollusks, arthropods, worms, namely, the epicotyledons, and some allocotyledons of Van Beneden, and finally, in the lowest, the echinoderms, the *coelenterata* and protozoa, that is, the majority of Van Beneden's allocotyledons.
11. {The importance of the work by Canestrini and others such as the physiologist and phrenologist Filippo Lussana, quoted by Canestrini, in regard to Paolucci's theories of animal language should not underestimated. Both elaborated on the notion that animals possess some form of language aimed at expressing the needs of social and psychic life. According to Canestrini, "speech based on sound and mimicry reveals clearly the state of animal feeling, and the will in animals demonstrates their intelligence. Language based on sound,

which is most common in mammals, in birds, in some reptiles, fish and insects, consists in sounds that express feeling, like the speech of the child in its first year, but it also includes sounds, vowels, noise and consonants, lasting more or less and repeated, whereas human speech is the result of the combination of articulated sounds in an orderly fashion according to precise laws. The sounds expressing feeling in animals are affine to the interjections of human speech, as these are but feelings dressed up as words" (Canestrini 1870, p. 23). Lussana conceived of language as the "manifestation of one's ideas, feelings and emotions to others; hence, [according to him] there are as many kinds of language as there are different means of communicating one's emotions and thoughts, namely, sounds, words, gesture, symbols, writing, and mimicry" (Lussana 1864, p. 274). Lussana's ground-breaking work on phrenology (*Lezioni di frenologia*) is fascinating. On the one hand he defends phrenology from the accusation of charlatanism and pseudoscience, while on the other he lays the philosophical foundations of scientific phrenology, cautioning philosophers against their prejudiced skepticism and insisting that empirical neuroscience and metaphysics should not be divorced from each other.}

12. Lubbock 1875. {Lubbock's book must have fascinated Paolucci. In the final part, Lubbock defends the view that civilization leads to greater happiness and pleasure and that the evolution and 'technology' of the senses contributed to a higher degree of happiness in civilized races than in savages. "There are, indeed, many who doubt whether happiness is increased by civilization, and who talk of the free and noble savage. But the true savage is neither free nor noble; he is a slave to his own wants, his own passions; imperfectly protected from the weather, he suffers from the cold by night and the heat of the sun by day; ignorant of agriculture, living by the chase, and improvident in success, hunger always stares him in the face, and often drives him to the dreadful alternative of cannibalism or death" (Lubbock 1869, p. 583). Indeed, "the possession of a new sense, or the improvement of am old one, is a fresh source of possible happiness; but how, it may be asked, does this affect the present question? There are no just grounds for expecting man, to be ever endued with a sixth sense; so far from being able to improve the organization of the eye or the ear, we cannot make one hair black or white, nor add one cubit to our stature. But on the other hand the invention of the telescope and microscope is equivalent in its results to an immense improvement of the eye, and opens up to us new worlds, fresh sources of interest and happiness. Again, we cannot alter the physical structure of the ear, but we can train it, we can invent new musical instruments, compose new melodies. The music of the savages is rude and melancholy; and thus, though the ear of man may not have appreciably altered, the pleasure which we may derive from it has been immensely increased" (Lubbock 1869, p. 586). Paolucci's work on birdsong and animal music challenges Lubbock's racist view that in the hierarchy of being civilized man must reign supreme as the solitary possessor of the most exquisite form of sonic sensitivity and aural happiness.}
13. *Fiji and the Fijians*. {Paolucci only quotes the main title of Thomas Williams's and James Calvert's two-volume book (Williams 1858 and Calvert 1858), without specific details. Cf. Williams 1858, p. 154, for Paolucci's remark.}
14. Astley 1746, p. 72. {"The officer who presents the cup turns his back, and gives it in that posture. They say this is done to prevent any charms or sorcery at that time" (ibid.).}

15. Speke 1863, p. 206. {I was unable to pinpoint the relevant information on the page referenced by Paolucci in Speke's book.}
16. Denham *Travels and discoveries in Africa*. Vol. II, p. 27, Vol. III, p. 15. {Cf. Denham, Clapperton, and Oudney, 1828, but I was unable to pinpoint the relevant information on the page of the second volume of the third edition that I consulted. I could not find a third volume.}
17. {Paolucci's idiomatic turn of phrase seems to indicate that he interpreted this gesture as if having a ridiculous or grotesque connotation, uncivilized, but this may or may not have been the case in Cook's original account. "I observed that they [i.e., the people of Mallicolo] could pronounce most of our words with great ease. They express their admiration by hissing like a goose" (Cook 1777, II, p. 36) On the other hand, Hook said: "…had we made a longer stay, we might soon have been upon good terms with this ape-like nation" (ibid., p. 34).}
18. {Here Paolucci used the Greek alphabet, which I could not reproduce.}

Bibliography

Astley, Thomas. 1746. *A new general collection of voyages and travels: Consisting of the most esteemed relations, which have hitherto published in any language: Comprehending everything remarkable in its kind, in Europe, Asia, Africa and America…* Vol. III. London: Printed for Thomas Astley, in Pater-Noster-Row.

Bagarotti, Giuseppe. 1857. *Nozioni fondamentali della scienza dell'umano intelletto*. Florence: Ferdinando Baccioni.

Beneden (van), Pierre-Joseph. 1853. *Anatomie comparée*. Brussels: A. Jamar.

Calvert, James. *Fiji and the Fijians. Vol. 2. Mission history*. Edited by George Stringer Rowe. London: Alexander Heylin.

Canestrini, Giovanni. 1870. *Origine dell' uomo*. Second edition. Milan: Gaetano Brigola.

Cook, James. 1777. *A voyage towards the South Pole, and round the world. Performed in His Majesty's ships the Resolution and Adventure, in the years 1772, 1773, 1774, and 1775. Written by James Cook, Commander of the Resolution. In two volumes.* The second edition. London: Printed for W. Strahan; and T. Cadell in the Strand.

Darwin, Charles. 1871a. *The descent of man, and selection in relation to sex*. 2 vols. London: John Murray.

———. 1871b. *L'origine dell'uomo e la scelta in rapporto col sesso*. Translated by Michele Lessona. Turin: Società L'Unione Tipografico Editrice.

Denham, [Dixon], Clapperton, Hugh, and Oudney, Walter. 1828. *Narrative of travels and discoveries in Northern and Central Africa: In the years 1822, 1823*. Third edition. 2 vols. London: John Murray.

Doron, Edith. 2008. *'Of things and thresholds': Exilic education, hospitality and the practice of community in the children's museum*. Masters of Modern Thought Dissertation. Aberdeen: University of Aberdeen.

Huber, Pierre. 1810. *Recherches sur les mœurs des fourmis indigènes*. Paris: Chez J. J. Pashoud.

Huxley, Aldous. 2009 [1954]. *The doors of perception*. New York, London, Toronto, Sidney, New Delhi, Auckland: HarperCollins. [1st ed. 1954, New York, Harper and Brothers].

Lifschitz, Avi. 2012. *Language and Enlightenment: The Berlin debates of the eighteenth century*. Oxford: Oxford University Press.

Lubbock, John. 1869. *Pre-historic times, as illustrated by ancient remains, and the manners and customs of modern savages*. London: Williams and Norgate.

———. 1875. *I tempi preistorici e l'origine dell'incivilimento*. Versione Italiana di Michele Lessona. Turin: Unione Tipografico-Editrice.

Lussana, Filippo. 1864. *Lezioni di frenologia*. Parma: Tipografia G. Ferrari.

Speke, John Hanning. 1863. *Journal of the discovery of the source of the Nile*. Edinburgh and London: William Blackwood and Sons.

Stoppani, Antonio. 1873. *Corso di geologia del professore Antonio Stoppani. Volume 2. Geologia stratigrafica*. Milan: G. Bernardoni e G. Brigola.

Süßmilch, Johann Peter. 1776. *Versuch eines Beweises, daß die erste Sprache ihren Ursprung nicht von Menschen, sondern allein vom Schöpfer erhalten habe*. Berlin: Zu finden im Buchladen der Realschule.

Wallace, Alfred Russel. 1876. *The geographical distribution of animals: With a study of the relations of living and extinct faunas as elucidating the past changes of the earth's surface*. 2 vols. London: Macmillan & Co.

Williams, Thomas. 1858. *Fiji and the Fijians. Vol. 1. The islands and their inhabitants*. Edited by George Stringer Rowe. London: Alexander Heylin.

Zanoni, Elena. 2014. *Scienza, patria e religione. Antonio Stoppani e la cultura italiana dell'Ottocento: Antonio Stoppani e la cultura italiana dell'Ottocento*. Milan: Franco Angeli.

PART ONE, CHAPTER TWO

The Mechanism of Sound in Autophonous Animals

In order to deepen our understanding of the mechanism of birdsong and the diverse biological functions that it serves, we must briefly discuss the organs of phonation in inferior animals and in some vertebrates. Once the anatomical differences between the organs of autophonous and heterophonous phonation (insofar as they are means of cognition) have been clarified, the enormous distance separating the two groups will become apparent. The examples I will discuss in this chapter will demonstrate that the function of vocalization in inferior animals is almost always the same, namely, sexual selection. It is the competition among the males for the possession of females (as Darwin suggested). Generally only the males sing. But they do not sing in the initial stages of their life when they are still sexually immature. They deploy all their vocal art during the short sexual period, towards which they are impelled by the miraculous love for the survival and progress of the species. In what follows, we will review the different zoological groups that are relevant for our purposes. I must point out that in regard to insect vocalizations I intend to refer to more or less harmonic sounds, not to the noises produced by many species—for instance, by means of sudden bodily movements, or by rubbing the serrated rims of certain thoracic and abdominal rings—which lack a sounding apparatus proper.

According to detailed observations of the furniture beetle (*Anobium tessellatum*) by Becker,[1] we know that also the female invites the male by hitting the ceiling of its nest with the forehead and the anterior margin of the prothorax. Becker repeatedly excited a female that he had kept with this signal while imitating the sound of the insect by beating the table with a knitting needle. He saw a couple having intercourse after the uncanny yet reciprocal seduction had been carried out by both lovers with passionate insistence. Again, among heterogeneous insects (according to Pierre André Latreille), there is a species of *mutilla* (*Mutilla Europaea*) the males of which can produce a stridulating noise whose function is not yet known.[2] In a large number of species of beetles both sexes can produce noise thanks to a rasping surface in proximity of a stiff protrusion. Sometimes it is the rasp that stridulates over the surface, sometimes it is the latter that engages the former. In the burying beetles (*Nicrophorus*) there are two rasps above the fifth abdominal ring that are excited by the internal rim of the wing cases (cf. Landois[3]). In a leaf beetle (*Clytra quadripunctata*) the array of rasping blades is located in the higher part of the abdomen. In some ground beetles (*Elaphrus uliginosus*, *Blethisa multipunctata*) and in the *Curculionidae* the abdominal rims rub the elytral scales. More or less the same happens in the water beetle (*Pelobius Hermanni*). In the earth-boring dung beetle (*Geotrupes stercorarius*) the abdomen creases a bladed protuberance in a slanted position on the thighs. More examples could be mentioned but the discussion so far suggests that these predispositions to produce embryonic noises will hardly allow any form of cognitive communication. They cannot be of much help in sexual selection in that both males and females make use of them. They represent a primitive stage in the evolution of the vocal apparatus of other insects, which have been perfected by the survival of the fittest. Let us now review the autophonous singing animals.

1. Insects

The poetess Annette von Droste-Hülshoff celebrates the ability of the best-known singing insects with the following verses. She describes the nature of their sounds beautifully, with exact similitude.[4]

From: "Die Lerche".	From: "The lark".
Da krimmelt, wimmelt es im Heidgezweige,	The bush is teaming with life,
Die Grille dreht geschwind das Beinchen um,	The cricket swiftly turns its leg,
Streicht an des Taues Kolophonium	And strokes the morning dew,
Und spielt so schäferlich die Liebesgeige.	Sweetly playing its love viola.
Ein tüchtiger Hornist, der Käfer, schnurrt;	A clever bugler, the beetle whirs.

Die Mücke schleift behend die Silberschwingen,
Dass heller der Triangel möge klingen;
Diskant und auch Tenor die Fliege surrt;
Und, immer mehrend ihren werten Gurt,
Die reiche Katze um des Leibes Mitten,
Ist als Bassist die Biene eingeschritten:
Schwerfällig hockend in der Blüte, rummeln
Das Kontraviolon die trägen Hummeln,—
[…]
So tausendstimmig stieg noch nie ein Chor,
Wie's musiziert aus grünem Heid hervor.

The mosquito sprightly rubs its silvery wings, so that it may brightly play the triangle;
Treble and tenor, the fly buzzes,
And, more and more increasing the gird in the middle of their abundant body, cats take action against the cello-playing bee, but squatting in blood the sluggish bumblebees hustle their double bass, […]
Thus, not even a thousand-voice choir would rise like the music from a green moorland.

We will distinguish between the sounds produced by the rapid vibrations of the wings from those made with a sonorous instrument built and used for this purpose. We will turn our attention to the latter sounds as the former are caused by the flight of insects all the time and involuntarily, though perhaps they might well serve the struggle for existence. Since these are not dependent on the power of the will, they cannot be regarded as means of cognition. For, cognitive acts requires the agent's will to express them in order for them to be understood by other agents. Nature has endowed many insects with vibrating organs, tensioned structures, and elastic reeds that to all intents and purposes function as string and wind instruments. The male that possesses one of these deploys it in general in the competition for sexual selection. The female chooses the most skillful player. However, one should not naively think that the winner won the competition because it was the superior virtuoso. The instrument could not have sounded but exactly the way it did. The winner turned out to be the best because it possessed the best strings and bow. In sharp contrast, we will see that this is not the case for birds, among whom the language of sound originates, transforms itself and is further refined in the same way as words and natural languages. In discussing insect vocalizations we will follow the divisions commonly accepted among zoologists.

1. *Diptera*

In flies (*Muscidae*) and mosquitoes (*Tipulidae*), aside from the sound produced by the fast vibration of wing, by the abdominal rings and hind-wings, we find a particular type of sound (studied by Landois) that can be produced by a genuine harmonic device that is connected with the four anterior thoracic stigmata or with the posterior. Many tracheal canals issue in a common chamber, the external aperture of which is the stigma. In this chamber there are many elastic reeds that are made to vibrate by air both when the insect inhales and when it exhales. But unfortunately we know little about these sounds as they are difficult to study. It is hard to say if and to what extent they might help the individual's struggle or natural selection. Patient and scrupulous research will hopefully uncover the mystery of that harsh sound, almost a lament, which mosquitoes produce when they molest us in search of the blood that nourishes them. We will also learn the cause of those concerts that are offered by the hideous mosquitoes of South America (*Culex molestus*, *trifurcatus*, *pulicarius*) when a huge swarm of males, frantically dancing, oscillates like a wondrous harmonic cloud, while the haughty females, known as *trompitas del diablo*, viciously bite the unlucky traveler. At any rate, these sounds are specifically made by autophonous animals and hence they are relatively confined to a low level of zoological organization.

2. *Orthoptera*

Among the invertebrate the *orthoptera* are the ablest singers in regard both to intensity and to the harmonic quality of the voice. At first glance it might seem strange that these insects appear to showcase their talent only for sexual competition. But their ability is limited to only one type of sound, a monotonous note, having always the same timbre and the same intensity. One sign could only serve to express one thing. Nature established that such thing should be love. The most delicate and pathetic member of this family is the elegant green cricket (*Locusta cantans*, Fabr.),[5] very commonly found in the countryside and even on the streets of our cities towards the end of summer. As is the costume among most animals the green cricket shares their instinct to choose the veil of night for the mystery of love, and towards sunset of a calm and sunny day one can hear the grand concert of these nifty creatures until the crack of dawn. The song is a sweet, light note, like a wavy whistle, a dying tremolo. The sound differs in accordance with the variations of the vibrating apparatus of the singers. If one listens to a multitude of them sing together, one perceives patterns and hence may have the illusion that they respond to one another's call. Here is an example in musical notation (Example 1).

FIGURE 2.1 Above: Example 1 (green cricket, two instances). Below: Example 2 (black field cricket). From: Paolucci 1878, p. 19. Note the complex rhythmic structure and the precise distance in pitch of one semitone between the two sounds (A sharp and B) by which Paolucci describes the song of the black field cricket. I am unclear why in the above example the first song is given in the upper portion of the staff while the second appears below, as the treble clef is the same (Caption by Paolo Palmieri).

The instrument of these harmonies is a delicate, tensioned membrane, circled by a conspicuous rim (the string), which is in the triangular dorsal part of the right elytron, underneath the left-hand side. In the inferior face, opposite the said membrane, there is a relief made by numerous twisting ribs, which is marked by many transversal lamellae (thus forming a bow). When the insect sings, the left elytron rubs against the right one, the bow causes friction on the string, and the membrane vibrates underneath the elytra, which constitute the resonance apparatus. When the rain season comes and it cools down a little, the green crickets stop singing, both their mission and the purpose of their song having been accomplished.

The music of another orthopteran, the black field cricket (*Gryllus campestris*), very common in our meadows in May, is much more shrieking, higher and sonorous. The serenade which this cricket indefatigably sings all night long close to the abode of its female companion, is like a stridulating whistle, accompanied by a tremolo, short, almost the same note in all individuals, which is repeated rapidly many times. I have attempted to transcribe it in music notation (Example 2, Figure 2.1). Again it is the toothed rib of the right elytron that produces the sound rubbing against another smooth, hard, and protruding rib in the left elytron. The only difference is the inverted position of string and bow.

Finally, also the Acrididae, known as grasshoppers, can produce some muffled sounds, which are broken, almost like noises, by rubbing the femurs against the elytra (in which there is inside a fine series of elastic teeth, shaped like spears). Again the vocal apparatus is a sort of violin. The animal has two of them and plays them alternatively. I suggest that the noise made by the common grasshopper (*Oedipoda migratoria*) in this way should be likened to the rustling of dried leaves, while that of another common grasshopper (*Tetrix subulata*) to the discontinuous hiss of a gushing liquid.

I have summarized and briefly described the music instruments of the *Orthoptera* with examples taken from indigenous insects belonging to the three families of *Locustidae*, *Achetidae*, and *Acrididae*. Many others of the same order sing their characteristic songs during the mating season, for instance, the English *Phasgonura viridissima*, the *Pneumora* of the Cape, the *Chlorocoelus tanana* from the Amazon, and so on. All of them vocalize always the same sound and express with it one intention.

3. *Homoptera*

Ever since the time of Anacreon that called the cicada "sweet singing muse"[6] and of the clever Xénarchus of Rhodes[7] who especially envied the cicada among the living for its having received a mute spouse as a present, this insect has been honored in the Parnassus. But nobody esteemed the cicada more than the erotic venerable old man from Teos.[8] Yet the useful animal has remained impervious to the curses of excessively sensitive people, and still sings its grandiose concertos every year until its last moments. Condemned to seventeen years of lethargy underneath the earth,[9] the cicada has but the fleeting period of one month to enjoy life *en plein air*, warmth, and light. Let us step into its shoes, then, and be charitable when it is seized by its canorous frenzy. The song of the common cicada (*Cicada plebeja*) and its close relatives is produced by a special apparatus located in the inferior portion of the body between the thorax and the abdomen. This device is covered by broad bony scales, which are free in the posterior part and towards the sides but fixed to the last thoracic ring in the anterior region. Underneath these scales there is a space, not too large, in which external air can penetrate freely when the abdomen is raised. The backstage of this vestibule (imagine seeing the animal upside down in dissection) is limited by two membranes on both sides and divided by a triangular process that is stiff and chitinous. The anterior membrane is yellow, delicate and soft like glove-leather. It is relaxed when the animal is at rest. The posterior one is iridescent, transparent, thin, and constantly tightened. It seals a large cavity having two lobes inside the abdomen, which is the sound box and which communicates with two large lateral spiracles. A complicated system of muscles operates the apparatus, which in the last analysis can be thought of as a drum. The tight skin is made to vibrate not so much by a flow of air as by a muscle connected to its center. The latter is the most recent opinion by M. G. Carlet.[10] Cicadas produce their stridulating song, which is sonorous, vibrant, interrupted at the beginning and in sync with the rising and lowering abdomen, then more continuous and eventually petering off, when the sonic apparatus and particularly the air chambers are emptied. Other species of cicadas that are fairly common among us have their own characteristic patterns. The flowering ash cicada (*Cicada orni*) has a low tone, weak, interrupted in long intervals, in contrast to the common cicada. On the other hand, the red cicada's song (*Cicada haematodes*) is more stridulous and almost continuous, or perhaps it is interrupted but just once.

2. Batrachians and Reptiles

As we move up in the tree of life beyond the vast class of insects, among which we have encountered scores of powerful musicians, adjunct members in the great orchestra of the world, we find the class of the mollusks. It is numerous and cosmopolitan, comprising, for instance, the speedy octopus that swims the oceans, the sluggish snails and shellfish, the minuscule sea angels (*Clionidae*), and other minor molluscoid organisms. But among its members mutism is the inescapable condition of their existence. As for the aquatic mollusks this condition is a natural consequence of the element in which they live, and thus in this regard we can class them together with fish. As for the terrestrial mollusks their social life is so limited that they scarcely need any means of communication. Since their vocal apparatus would be useless, it is indeed missing. Sometimes we find otoliths, like a sketchy auditory organ, almost a first step towards hearing in the developmental biology of these organisms. It seems that these animals are unable to understand or make themselves understood phonetically even though they might hear.

As we move upwards in the zoological series we find the superior animals, namely, the vertebrates. We will not consider fish as they have no voice. The batrachians and reptiles are the first and simplest class of singing vertebrates, which should not be confused with the insects in that, as we have seen, the latter make sound only by playing string-like instruments.

Some reptiles and batrachians produce voices, shouting, sibilant and hissing sounds with the same mechanism as that with which birds sing, mammals shout, and human beings speak. In other words, it can always be analyzed as having a larynx in connection with lungs that work like bellows (as in reptiles), or in communication with air chambers; for instance, pockets located underneath or besides the head (as in anurans), pneumatic reservoirs that, as we shall demonstrate, function in the same way as those found in birds. The vocal apparatus in reptiles and batrachians has some features that are required for sound modulation. Indeed, the form of this apparatus, its anatomical structure, is in outline the same as that of birds and mammals. But it is imperfect, which explains why only birds and mammals are able to modulate sound. Observation has revealed that the blowing of turtles, the hissing of some snakes, or the shrieks of certain crocodiles, are always identical for that species, that is, the individual animal can only repeat the same sound. Therefore, these animals are autophonous like insects, even though their sounds are produced by the larynx, which means that the animal in reality does not 'play' but rather 'sings', precisely like the heterophonous animals. As the individual animals among reptiles and batrachians cannot produce a variety of sounds, although they are endowed with a larynx structure, we must draw the conclusion that they represent a transition from inferior animals, necessarily autophonous, to superior ones, which are heterophonous both in terms of structure and cognitive behavior. The development of the vocal apparatus of these variable-temperature vertebrates would have been arrested. According to Darwin, further development might have been impeded by non-use, owing to the limited cognitive abilities of these animals.

At any rate, the sounds produced by the reptiles that have some voice are scarce, occasional, and modest, in accordance with the low-level of these animals in the zoological series. Brehm claims that some turtles make sounds like a hiss when they are subject to high levels of stress, and the same is true of snakes in regard to whose hissing, though, too many legends have been repeated. This means that their sounds are produced under such extraordinary circumstances as to make it unlikely that they might serve as a basis for language. The bestial voice of crocodiles appears to be more intelligible. Its meaning is perfectly understood by those who come up against that monstrous descendant from long extinct species. Also the gecko (*Hemidactylus verruculatus*), when it pops up at night, hunting down insects inside the furniture of our houses, makes acute and bright sounds, which are very characteristic, like a *cich cich*.

So far we have not talked about sexual language in reptiles, a subject which we discussed in detail in the case of insects. As far as I know, neither the *Cheloniidae*, the *Ophidia*, nor the saurians, produce any love sound whatever related to or during sexual competition or mating. Among the batrachians, on the other hand, the singers use and abuse their vocal organs during the love season. Only the males sing. They sing until they mate with the female, which objective is the leitmotiv of their songs. Among the autophonous vertebrate they correspond to the grasshopper, the cricket, and cicadas. Everybody who has lived in a town close to swamps and stagnant waters will have noticed, with pleasure or annoyance, the persistent music of frogs, or their aggressive statements, or the melancholic and almost gloomy voice of the toad. Those who have a passion for delicate breezes and the aroma of young leaves must have fond memories of these concerts. Already in March they listen with great pleasure to the early, tentative notes, sonorous harbingers of the annual resurrection that nature experiences in temperate climates. But why do those timid dwellers of ponds, which remain mute and sleepy throughout winter, reawaken so sprightly in spring and so willing to sing? The rise in temperature, a supremely important factor for all biological life, is most effective on them as well as on animals and plants. It powerfully energizes their body so that they might accomplish their mission, namely, the reproduction of individuals. Indeed, it is during sexual reproduction that the animal must deploy all its faculties. Hence the development of self-defense organs, the luxuriant display of colors, and the amorous songs, all employed in the sexual game.

I think that the song of many batrachians is a subtle strategy for sexual competition. I consider it a genuine love language. There are two facts that support my claim. First, these songs are produced only during mating season. Second, the females do not sing, for they are under the selective pressure of males. Furthermore, batrachians rarely sing alone, so that the concert appears to me to be another argument for its sexual purpose. The concert of the males is like a prize competition. The female awards the contestant that sings the best song the coveted prize. I have come to this conclusion by analogy, while reflecting on the same phenomenon that, as we shall see, is much more obvious in birds. The theory that the song of batrachians is a sexual language was also reinforced by the observation that the green toad (*Bufo viridis*) follows the same pattern of behavior as the locust. A few males close to one another so as to understand one another produce their characteristic song, which is a blunt whistle, yet sonorous and gurgling, during the same period of time. It could be transcribed as follows (Figure 2.2, above).

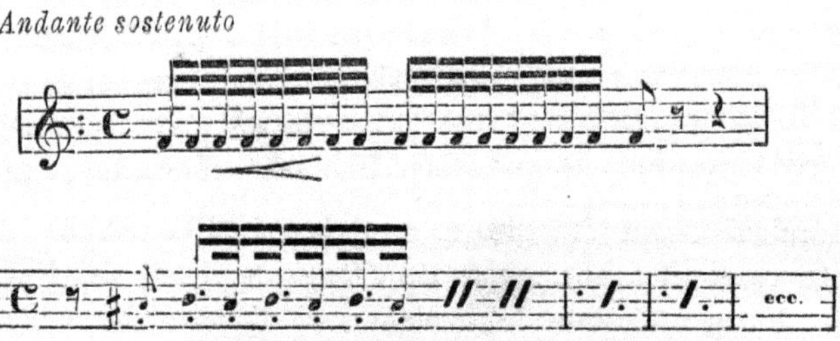

FIGURE 2.2 Above: The song of *Bufo viridis*. Below: The shrill noise of *Hyla viridis*. From Paolucci 1878, pp. 26–27. Caption by Paolo Palmieri.

However, it rarely happens that two singers sing together. Most of the time one begins when the other has finished. The ensemble proceeds by successive solos. It should be noted that the note of the song is not the same in all individuals, and the difference in pitch can be significant, even two or three semitones, yet the intensity of the voice remains the same. So the victory will be awarded to the toad that sings the most gracious note not the loudest. This suggests that each individual wishes to be appreciated as a soloist rather than as the member of an ensemble.

A similar though more confused pattern of behavior can be observed among edible frogs (*Rana esculenta*). But the pitch of the tone and/or the chorus is the same or almost the same for all singers. Thus it seems that the competition is based more on intensity rather than tone. As Brehm noted,[11] when a male starts the concert, a veritable racket of untrained singers follows suit, an intolerable, confused noise caused by the chorus of multiple notes too close to one another. "Brekeké!" exclaims the concert master, and all of a sudden… "brekeke—brekeke, brekeke—koah—brekeke, brekeke—brekeke, quarch brekeke—brekeke, brekeke, brekeke—brekeke, brekeke, brekeke, brekeke—koah, koah, tuu—brekeke, tuu—brekeke—brekeke…", answers the chorus of the singers.[12]

In the green tree frog (*Hyla viridis*) the structure of the larynx is so similar in all male individuals that they all sing the same note having the same pitch and the same timbre. It is that well-known, high-pitched, shrill *tarac*, *tarac*, *tarac*, which might be rendered musically as the second example notated in Figure 2.2. In this case, the competition cannot take place through the alternating voices as they perfectly resemble one another. Only the strongest singer will be heard, and hence the competition is based on the simultaneity of voices. Indeed, the song of the green tree frogs is a magnificent chorus, sometimes so perfectly in unison that the simultaneity of hundreds of garrulous *tarac*, *tarac*, *tarac*, *tarac*, is distinctly perceived. It is amazing to hear the sonorous burst of countless males that respond to the first *tarac* without a single individual answering the cue with ever so slight a delay. Special circumstances such as weather and the degree of humidity might favor sexual appetite in many anurans and hence might be related to singing. In the case of the green tree frog I have observed that any loud sound may trigger singing, as I will report in detail presently.

It was a tranquil night in May. The garrulous dwellers of the ponds around Ancona's fortresses were all quiet. A trumpet sounded the silence for the soldiers. But at the same time, as it turned out, it gave the signal for countless frogs that at once burst out in a ripping sound. By chance I observed the same fact a second time. I am inclined to believe that the sound may have excited the frogs to singing, perhaps because it was perceived as a cue given by one of them.

What I have said so far concerning the anurans I mentioned is also true for other batrachians in tropical regions, which often sing together very loudly. The Brazilian yellow frog (*Hyla luteola*) sings, according to the Prince of Wied,[13] in a very similar way to our green frogs. Travelers and trusted naturalists attribute to the American bullfrog (*Rana mugiens*) a powerful voice. The diverse voices of the *Ceratophrys Bojei*, of the *Pyxicephalus adspersus,*[14] and of a few more *Hylae*[15] and batrachians have been catalogued. However, observations are still deficient and vague, and therefore we can hardly draw definitive conclusions as to our main subject.

In sum, autophonous singing in insects is made possible by instrumental apparatus, while in batrachians by the vocal apparatus. The purpose of the songs in these animals is narrow, and so far as we can say it is oriented towards sexuality.

Notes

1. {Possibly Ludwig Becker. Cf. Ninnis 2011, p. 313.}
2. According to Goureau and Darwin 1871, p. 264. {(Claude Charles) Goureau, quoted by Darwin, ibid.}
3. {See Landois 1867.}
4. {The German text is from Brehm 1900, p. 17. Paolucci quotes the poem in the (rather poor) Italian version given in Brehm 1875, p. 17). The English translation from von Droste-Hülshoff's lovely original is mine (for what is worth).}
5. {Paolucci gives the Latin name followed by the abbreviation Fabr., which stands for Johann Christian Fabricius, Danish entomologist and author of many works on insects. Cf. Fabricius 1775, for instance.}
6. {Anacreon is quoted in Italian from the translation by Bartolomeo Corsini (see Corsini 1672, pp. 51–53).}

7. {I think Paolucci has in mind Xénarchus the comedian of the fourth century BCE though I am not sure as to the source from which Paolucci must have taken the information that Xénarchus was from Rhodes. I believe it was the sixth volume of the Italian edition of Brehm's *Life of animals* (Brehm 1875, pp. 559–562), where it is reported that Xénarchus from Rhodes wrote that "Ha la cicala sorte avventurosa,/ Perché senza favella è la sua sposa" (ibid., p. 560).}
8. {…namely, Anacreon.}
9. Canestrini 1870, p. 132. {Paolucci erroneously cites Canestrini from p. 130, I have corrected the reference. Interestingly, unlike Paolucci Canestrini qualifies the song of cicadas as monotonous, boring and annoying, "…monotono, ingrato e fastidioso" (ibid., p. 204).}
10. Carlet 1877.
11. {Although in a footnote Paolucci cites Brehm's fifth volume giving the title in German (without other bibliographic details) and referencing page 428, I think he actually read the Italian edition (Brehm 1872, p. 428) because in the fifth volume of the German edition (Brehm 1869) the page numbers of the pertinent paragraphs on edible frogs are different (ibid., pp. 378–380). The second German edition of Brehm's work, which Paolucci might have consulted, most likely appeared too late for him (1878, same year as his memoir), and at any rate both the relevant volume and the pagination are once again different. Brehm reported that he had enjoyed the concert of the frogs in the countryside and that he appreciated both the happiness and harmony in their songs, though by most people, he acknowledged, their concert was perceived as annoying. Brehm did not use the language of noise with which Paolucci qualifies the sound of frogs. I believe that here Paolucci is at the very least imprecise and arbitrarily attributes to Brehm negative views.}
12. {In both Brehm 1872, p. 428, and 1869, p. 379, the transcription of the song is slightly different than the version reported by Paolucci, who has 'koah' instead of 'koax' used by Brehm and the faithful Italian translators. It is unclear whether Paolucci misread the source or intended to insert a correction based on his own observations. The latter case is more likely, I believe, given that he almost certainly listened to the songs of these batrachians as he gives us to understand in the subsequent discussion.}
13. {Most likely, Maximilian, Prince of Wied (1782–1867), see Vanzolini and Myers 2015.}
14. Brehm 1872, p. 440. {I have corrected the reference again, quoting the Italian edition of Brehm's work.}
15. Darwin 1871, p. 321.

Bibliography

Brehm, Alfred. 1869. *Illustrirtes Thierleben*. Fünfter Band. Hildburghausen: Verlag des Bibliographischen Instituts.

———. 1872. *La vita degli animali, descrizione generale del regno animale. Traduzione Italiana dei Professori G. Branca, S. Travella, e altri naturalisti. Riveduta da Michele Lessona e Tommaso Salvadori. Volume Quinto.* Turin: Unione Tipografico Editrice Torinese.

———. 1875. *La vita degli animali, descrizione generale del regno animale. Traduzione Italiana dei Professori G. Branca, S. Travella, e altri naturalisti. Riveduta da Michele Lessona e Tommaso Salvadori. Volume Sesto.* Turin: Unione Tipografico Editrice Torinese.

———. 1900. *Tierleben; allgemeine kunde des thierreichs. Mit 1910 abbildungen im text, 11 karten und 180 tafeln in farbendruck und holzschnitt. Dritte, gänzlich neubearbeiten Auflage. Insekten.* Leipzing und Wien: Bibliographisches Institut.

Canestrini, Giovanni. 1870. *Compendio di zoologia ed anatomia comparata. Parte II: Molluschi ed Artropodi.* Milan: Gaetano Brigola.

Carlet, M. Gaston. 1877. "Le chant de la cigale". *La Revue scientifique de la France et de l'Étranger. Revue des courses scientifiques.* 2e série, 7e année, 22, pp. 516–519.

Corsini, Bartolomeo. 1672. *Anacreonte poeta greco. Tradotto in verso toscano, da Bartolommeo Corsini.* Paris: n.p.

Darwin, Charles. 1871. *L'origine dell'uomo e la scelta in rapporto col sesso.* Translated by Michele Lessona. Turin: Società L'Unione Tipografico Editrice.

Fabricius, Johann Christian. 1775. *Systema entomologiae: sistens insectorum classes, ordines, genera, species, adiectis synonymis, locis, descriptionibus, observationibus.* Flensburg et Leipzig: In Officina Libraria Kortii.

Landois, Hermann. 1867. *Die Ton- und Stimmapparate der Insecten: in anatomisch-physiologischer und akustischer Beziehung.* Leipzig: Verlag von Wilhelm Engelmann.

Ninnis, Elizabeth. 2011. "The art of the Victorian exploring expedition". In: E. B. Joyce & D. A. McCann [Eds.], *Burke and Wills: The scientific legacy of the Victorian exploring expedition.* CSIRO Publishing, Collingwood, Australia, pp. 312–327.

Paolucci, Luigi. 1878. *Il canto degli uccelli. Note di fisiologia e biologia zoologica in rapporto alla scelta sessuale e alla lotta per l'esistenza.* Milan: G. Bernardoni. Retrievable online at: http://dfgviewer.de/show/?set[mets]=http%3A%2F%2Fwww.zvdd.de%2Fdms%2Fmetsresolver%2F%3FPPN%3DPPN641658559&set[image]=1

Vanzolini, Paulo E., and Myers, Charles, W. 2015. "The herpetological collection of Maximilian, Prince of Wied (1782–1867), with special reference to Brazilian materials". *Bulletin of the American Museum of Natural History Number* 395, pp. 1–155.

PART ONE, CHAPTER THREE

The Mechanism of Sound in Heterophonous Animals

We recall that we labeled 'heterophonous' the animal that is capable of producing vocalizations under different circumstances and for different purposes. In the last chapter we dealt with the zoological classes belonging in the larger division of the autophonous animals. We will now turn to mammals and birds, which are all capable of heterophonous vocalizations. I believe that this is the privilege of superior animals, and more precisely of the vertebrates with constant temperature. In general it might be stated that the animal's capacity for sound modulation and vocalization is directly proportional to the differentiation of the organs, that is, to the perfection and the rank in the zoological tree. However, this law does not seem to be valid when one compares and contrasts the capacity for vocalization in the minor groups, genera, and families of warm blooded animals. On the other hand, the difficulties almost disappear is one takes into account certain biological facts.

In the remainder of this work I will try to demonstrate that birds have a much more developed capacity for sound modulation than mammals. This would contradict the above mentioned law even if I qualified my claim by asserting that it is only true in a first approximation in relation to the entire spectrum of animal life. But in fact there is another empirical regularity in biology that furnishes the reason why birds make a more extensive and more diverse use of vocalization. In animals I have often observed the following pattern, which I would call 'the law of functional substitution'. It is the fact that an animal may take an action for a purpose related to social life by which it replaces a different action undertaken by other animals for the same purpose. In other words, when an animal is unable for biological reasons to take two or more actions related to social life with differently appropriate means, it will make use of the same means of expression that under such circumstances will replace the differentiated actions of which a more perfect animal is capable. It is beyond the scope of the present work to argue more cogently in support of the claim I just made. The following considerations will have to suffice for the time being.

Instances of functional substitution will tend to occur in animals all the more frequently the less differentiated their organs are. For, under such circumstances, an organ will have to perform more than one function,[1] and the animal will need to accomplish more than one goal in social life by means of one action only. For example, because of both a paucity of organs and less complex cognitive abilities, certain animals will have to perform two functions with one organ, or start two social relations with the same signs, as opposed to other animals, which might be endowed with different organs and hence be able to communicate with different signs. At a lower level of biological organization animals substituted one single organ, or single action, for a plurality of organs and action, thus instantiating a compendium of their own physiological and biological attitudes. The same happens in birds in comparison with mammals as to their heterophonous capabilities. Often the birds make use of the voice for many more purposes than mammals. Whereas a bird always makes use of the voice for a certain purpose, a mammal is able to undertake different patterns of behavior for different purposes. In general, the biological objectives that a mammal very often accomplishes by means of smell, sight, hearing, muscular power and defensive mechanisms adapted to terrestrial life, are attained by birds solely by means of the voice. One reason why birds sing for multiple purposes might be that they are predisposed to flying, a fact that makes it easier and more natural for them to sing, in sharp contrast to mammals, which might find it more difficult if not impossible. In subsequent chapters, we will see that birdsong can be interpreted differently when viewed as animal language. However, when we say that birds vocalize more than mammals because they are structurally inferior to mammals, the explanation cannot be considered as applicable to the whole animal kingdom, for otherwise it would follow that reptiles would vocalize more than birds, batrachians more that reptiles, and so on. The explanation is valid only to the extent that a sufficient level of cognitive ability appears in a certain animal. Although they are inferior to mammals, birds are endowed with sufficiently powerful brain structures that allow them to communicate certain voluntary patterns and make manifest their passions.

We already noted that in general we find a more complex attitude towards the articulation of sounds in superior animals. Now, humans are the living beings that can articulate sound in the most exquisite manner. This is because of their anatomical structure, setting aside for a moment their intelligence, much more advanced than that which we find even in their closest relatives, the great apes. The wonderful and delicate capacity for articulation, which shines through the myriad human languages, is due to the shape and disposition of the lips, of the tongue and teeth, of the palatal vault and veil, of the oral cavity, of the nasal cavity and the pharynx. The larynx and vocal cords would barely have made it possible for humans to express the vowels, which means that they would only have been capable of leveraging phonetic materials as much, or as little, as any other mammal. Even though we could imagine another mammal endowed with the same intelligence as humans, we could not but conclude that it would still lack the complex anatomical structures required for speech. Physiology teaches us with scientific rigor the precise position that the lips, the tongue, the palatal veil, and the pharynx must assume for the exact emission of consonants, according as they are lingual, dental, labial, nasal, guttural, and so on. Now, it is true that certain animals such as parrots are able to articulate words and even entire phrases more or less convincingly. But the reason for this peculiarity is to be found in the form of their oral cavity, of their superior larynx, and especially in the muscular structure of the tongue, which resembles that of humans.

The mechanism of phonation is identical in all mammals. During inhalation the lungs and vesicles fill with air, and during exhalation they work as bellows pressurizing an air column that impart a more or less frequent motion to the vocal cords and the larynx. The oral cavity responds like a bell, or rather as a resonance apparatus. The pitch of the note, or voice emitted depends on the number of vibrations produced in an interval of time by the vocal cords, its power, or force, on the amplitude of the sound wave, and finally the timbre, or tone color (as the Germans aptly call it, namely, *Tonfarbe*), on the form of the vibrations.[2] According to Helmholtz, every uttered vowel has a different color or timbre. So, if one pronounces or sings *a e i o u*, one hears the note of a flute, or a violin, or an oboe, or a trumpet, or a clarinet, respectively. The vocal apparatus of a mammal, as it turns out, is such a vastly more complex musical instrument than any artificial one yet invented by mankind. It can sound like an entire orchestra, all the more numerous according to the number of vowels that the animal can produce. In fact, although there are only five considered to be fundamental, typical, indeed essential, yet the number of perceived vowel-like sounds, in their infinite nuances, greatly exceeds that figure.

Thus, I wanted to evoke the wonders of the vocal apparatus in mammals so as to be able to draw an analogy with the mechanism of phonation in birds. In these lofty and gentle creatures, which appear to have been shaped by nature following the aesthetic canons of musical beauty, mutism is a rare exception. Dissection has shown that their singing apparatus is a formidable musical instrument. Very often the entire body of the bird plays the instrument with great passion. The subtle patterns of song follow the whims of the composer with grace, rapidly passing from the feeble adagio of a symphony to the allegro of a dance, from the sprightly thrill to a long-sustained note, from the raucous bowing of the violin to the supple movements of the flute.[3] Like in mammals the vocal apparatus of birds is part of the respiratory system. In essence, it consists of a larynx made of cartilaginous elements which are linked to one another for fine-motor articulation. Unlike in mammals the larynx of birds is not located close to the superior extremity of the trachea, where instead there is a simple longitudinal aperture without epiglottis, which can barely influence their vocal power. Thus, in birds, because of the different anatomical structures, the oral cavity cannot work as resonance chamber or bell, and indeed it can remain partly open when the animal sings. The organ of phonation is situated close to the bronchial bifurcation, in the vicinity of the last tracheal ring, next to the pulmonary extremity of the trachea, which in turns operates as a sound box for the sonorous vibrations. Consequently, their trachea functions acoustically like the oral cavity in humans and superior animals. To this purpose it modifies the voice by becoming narrower or larger, or shorter, according to the timbre of the note, that is, its vowel-like or instrumental quality. In 1784, Michele Girardi, a keen observer and skilful dissector, recognized that there are two or four fixed muscles in the trachea that, by alternately moving it closer to, or away from, the bronchia, dramatically influence the intensity of the voice.[4] His observations have been confirmed by recent anatomists. To all intents and purposes, the trachea in birds must be considered as a sophisticated controller of the sound emitted by the larynx.

As we have seen, the larynx is an expansion of the last cartilaginous ring of the inferior part of the trachea. The structure of the larynx is more or less complex in different species, in accordance with their different capacity for phonation. Canestrini succinctly describes it as follows.

> Along the dividing line of the trachea, between the two bronchia, there exists a bony lamina that is extended downwards, in both the anterior and posterior sides, and which tensions a membrane. The latter is attached to the above mentioned bony lamina, and is called 'interior eardrum-like membrane' (*membrana tympaniformis interna*). Between the last tracheal ring and first bronchial ring there is a second membrane which, when the rings come close to each other, bends towards the interior side, moving closer to the eardrum-like membrane. This second membrane is the exterior eardrum-like membrane. These two membranes function like vocal cords. In singing

> birds there is a third membrane, originating in the bony lamina. It is a prolongation of the interior eardrum-like membrane. It is called 'semi-lunar membrane' (*membrana semilunaris*). Sometimes the inferior portion of the trachea forms a cavity that works like a sound box.[5]

Furthermore, Canestrini distinguishes three types of inferior larynx. First, the larynx formed solely by the trachea, without bronchia, which is controlled by one to three pairs of muscles located besides it. Second, the larynx formed by both the trachea and the bronchia, again controlled by one to three pairs of muscles located besides it. Third, the larynx formed by both the trachea and the bronchia, and controlled by two to five pairs of muscles, located both in front of it and behind it. Accordingly, in relation to the different structure of the inferior larynx, birds might be classified as *tracheaphonous*, *vocalizers*, and *singers*. However, I think that, although there is no doubt as to the importance of the inferior larynx in the mechanism of phonation, other organs must be taken into account that play a role in modulating the voice of birds. These are the trachea, which we touched on already, and the so-called *air pockets*, which I will presently comment on very briefly.

In general, birds are superior to mammals in many vital functions. Their rapidity of movement, expressing joyful impetuosity, their exquisite senses, their energetic passions are extraordinary. The hallmark of their powerful character appears in their furious struggles, in their enjoyments and loves. Such an effervescence of life is related to the fast oxidation of their tissues since the average temperature of their body is believed to be five degrees higher than in humans.[6] These phenomena can be explained by the fantastic pulmonary activity, which occurs in birds, because the pulmonary tree does not terminate, as in mammals, in many closed vesicles. Instead, it takes the form of a labyrinth network, communicating with large reservoirs of air, which allow the circulation of air even in bones and feathers. Certain naturalists compared these reservoirs with the pulmonary vesicles of mammals. Others such as Ph. C. Sappey, in a most accurate monograph, think that they are specific organs, analogous to the special structure of vertebrates that need them.[7] Sappey considers nine reservoirs, as follows.[8] 1) The thorax reservoir, in the anterior part of the thorax. 2) Two cervical reservoirs, at the base of the neck. 3) Two diaphragmatic anterior reservoirs, placed between the two diaphragms.[9] 4) Two diaphragmatic posterior reservoirs, situated between the two diaphragms, behind the anterior ones. 5) Two abdominal reservoirs, close to the inferior face of the abdominal wall. The first reservoir is asymmetric, the others are symmetric. As for their anatomical relation to the lungs, one must remember that the thoracic reservoir and the two cervical ones are in front of the lung, the abdominal one are behind, and the diaphragmatic ones are below. Thus, they can be all divided in anterior, posterior, and intermediate (Figure 3.1).

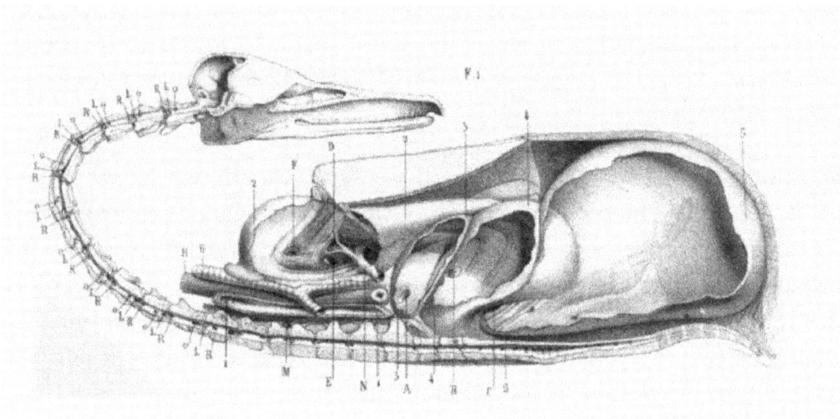

FIGURE 3.1 From: Sappey 1847, *Figure 1, Engraving III* (explained in detail in ibid., p. 89). A lateral view of the reservoirs of air open from the interior face as illustrated by Sappey. Caption and figure by Paolo Palmieri.

These air pockets have great influence on the weight of the body, on the balance of the animal, and on the overall mechanism of forces. In addition, among other noteworthy uses, it is most important to realize that they exert a powerful action on singing. Otherwise, how could the surprising intensity of song be explained in small birds, such as the nightingale, the robin, the canary, the common chiffchaff, solely by the air capacity of the lungs, which are relatively small, without having recourse to the mechanism of efficient and powerful bellows? The experiments carried out by Sappey demonstrate that the intermediate reservoirs are those which push the air towards the larynx during respiration, while the anterior and posterior ones fill with air when the bird sings, thus weakening rather than strengthening the power of the voice. The art of modulating the voice, therefore, whose perfection we admire in so many species in our forests, is plausibly explained, in my opinion, through the mechanism of the anterior and posterior air pockets, which act in counterpoint to the intermediate ones. The lungs in birds are like bellows of organs, while the diaphragmatic reservoirs represent the reinforcing pedal, and the others work like the damper pedal.

Let us also remember that certain birds that produce special vocalizations during the mating season are endowed with resonant cavities on the side of the head and of the neck, which fill up with air while the inferior trachea vibrates. The prairie hen (*Cupidonia Americana*) develops two yellow-orange vesicles on the sides of the neck and of the head, by which it produces a special noise, similar to the sound of a drum.[10] Audubon's experiments demonstrate that these pockets are a powerful instrument for singing. The Amazonian umbrella-bird (*Cephalopterus ornatus*) wears on the neck a long protuberance covered in short scale-like feathers, which is not only ornamental but works as a resonant apparatus. Indeed, according to Bates, it is connected with the trachea.[11] Numerous species of great bustard (*Otis tarda*), such as the common European one, according to the studies of A. Newton, Cullen, Flower and Murie, carry a bag on the throat that during the mating season helps especially the males to produce certain raucous sounds.[12]

We needed to touch briefly on comparative anatomy and physiology to recall stock facts that have been established by science, and which will be useful for the subsequent discussion even though they appear to be underestimated or looked at suspiciously, or occasionally not taken seriously, by naturalists. If natural selection has worked such wonders in building a complex vocal apparatus in the organism of birds, it must have a superior purpose than the expression of rudimentary instincts. It cannot be there to work as pastime, or to cause pointless labor, or indeed as a secret virtue destined to remain incomprehensible to the birds themselves. Wallace's, Darwin's, and Haeckel's studies on animal and plant biology, to which contributions have been made also by Italian scientists, have revealed the purposes of many forms and functions in the general economy of organisms. I believe it would be in contradiction to the fundamental laws of evolution to acknowledge the complicated structure of birdsong without also acknowledging a proportional importance to its function.

Notes

1. Canestrini 1869, p. 11.
2. According to the experiments by Helmholtz. {Paolucci has in mind the ground-breaking researches on the sensations of hearing by Hermann von Helmholtz. The study of the neurophysiology of the sense of hearing in humans and mammals especially became a scientific field in its own (modern) right around the middle of the nineteenth century. The foundations were laid by physicists such as Georg Ohm and Hermann von Helmholtz and by the physiologist Alfonso Corti. Ohm and Helmholtz realized that importing the mathematical techniques of Fourier analysis into the field of hearing would pave the way for a creative understanding of how we perceive the world of sounds (recall that Joseph Fourier was the French mathematician who at the beginning of the nineteenth century worked out the pioneering algorithms that allow the mathematical representation of periodic phenomena by infinite systems of sines and cosines functions). But it was Helmholtz that coupled that mathematical insight with the fundamental discoveries in the hearing physiology of mammals by Alfonso Corti (in whose honor the complex of organs of the mammal inner ear has since been named *organ of Corti*, cf. Corti 1851). Thus, Helmholtz formulated a theory of hearing whose general thrust dominated the field for many decades (and still does so today, to a very large extent; on which debatable, puzzling question, may I pause

and take the liberty of referring the reader to my Palmieri 2012). See Helmholtz 1863 (first edition), 1877 (fourth and definitive edition, and the excellent English translation (or perhaps *domesticated* re-making?), 1885. The crosspollination of Helmholtzean fantasies on the sensation of pitch, timbre, and the perception of music with the empirical investigation of animal languages and a penchant for linguistic speculation launched Paolucci into an adventurous exploration. He created an exotic, even extravagant imaginative framework. In that milieu of blossoming, colorful ideas, his hermeneutic of birdsong opened up expansive philosophical vistas, far beyond the intellectual narrowness of European positivistic science. The same will appear true of Helmholtz's work on music and natural philosophy if and when it is read, at last, in the spirit of anarchic freedom of thought in which it was conceived, a reading which will bring to the fore the disciplinary and political prejudices that have killed off academic freedom in our contemporary (dis)educational systems.}

3. {Paolucci uses the words 'complicata armonia' (which I have rendered as 'subtle patterns of song', avoiding the cognate word 'harmony', which would likely be heard as meaning 'the sounding together of many voices') to evoke the musicality and structural complexity of birdsong though it is obviously monophonic (except in choruses, on which Paolucci will say more further on). The word 'armonia' points to complex layers of meaning, deposited in the literary, scientific, and philosophical traditions of the last few centuries of European culture. The important nineteenth-century Italian dictionary known as 'Tommaseo-Bellini' (see a practical online edition at: http://www.tommaseobellini.it/#/), emphasizes the meaning of order resulting from the agreement of diverse elements, though it also points out that not every order constitutes 'armonia'. A very deep layer of meaning, which might escape the casual reader, implies certain philosophical assumptions as to the structure of perception, which are rooted in the early modern transformations of the classical Aristotelian theory of perception. We find a description of this layer and its substrata in the widely popular *Discorsi accademici* of the naturalist and classicist Anton Maria Salvini (1653–1729) (also quoted in Tommaseo-Bellini, *ad vocem* armonia). "Nella soavità ancora de' suoni e nell'armonia non solo di quelli, ma degli odori, dei sapori, e delle tangibili qualità e differenze, come del raro, del denso, del cedente, del solido, del fluido, la bellezza al comun sentimento (giudice di tutti i particolari sentimenti) si presenta, e di sè l'innamora" (Salvini 1735, p. 119). Here Salvini notices that it is the 'common sense', presiding over the specialized five senses, that is the recipient, or perhaps the creator, of 'armonia', the nexus of orderly structures that originate the cognition of beauty. Salvini saw clearly that the definition of harmony presupposes the philosophical dilemma of the reality of perceived qualities, namely, the question of their status as entities existing in the world as opposed to purely mental realities. He pointedly asked the question: "Se la bellezza sia una forma risultante dalla proporzione delle parti, o pure abbia la sua esistenza nella opinione di chi la rimira; farci nello stesso tempo così alla sfugita considerare quella gran question filosofica, e tra grandi autori dibattuta: Se le qualità sieno cosa reale, e che senza che niuno le riceva, o le consideri, nella natura, e nel suggetto medesimo abbiano sua residenza, o pure l'esser loro riconscano solo da' sentimenti nostri, e dalla nostra considerazione" (ibid., p. 119). The most important formulation of the question was given by Galileo Galilei (also discussed by Salvini) in his celebrated *Il Saggiatore* (1623, pp. 196 ff.), in which the Italian astronomer

laid the philosophical foundations of the modern debate on the perception of primary and secondary qualities that has challenged Western philosophers until the present time.}

4. Girardi 1784, p. 732. {Michele Girardi, anatomist and medical doctor, had cautioned that the dissection of the organs of phonation and respiration in birds is extremely difficult, and that there is great structural variation among individuals of the same species. He also pointed out that he had often observed, even in a single individual, a dissymmetry in the number and location of air vesicles between right-hand side and left-hand side of the bird. Cf. ibid, pp. 732–733, and see Porro 2001, on Girardi's scientific career.}
5. Canestrini 1869, pp. 159 ff.
6. Menault 1872 [1868]. {The first edition, which I have consulted and listed in the bibliography, was published in 1868. Menault gave a fascinating account of animal intelligence, placing himself in the venerable tradition of mystics and thinkers such as, Lactantius, Saint Francis of Assisi, and Michel de Montaigne, and insisting that "to a certain extent animals possess the same cognitive faculties as humans, and that our inferior brothers, as Saint Francis of Assisi called them, preceded us on the earth and have been our first teachers. We believe, together with Montaigne, Réaumur, la Fontaine, Georges Leroy, Frédéric Cuvier, etc., in the intelligence of animals [*l'intelligence des bêtes*]; we acknowledge, with Cabanis, that there are subtle relationships between biological organization and intellectual faculties, and with Gall that intelligence has its seat principally in the brain" (ibid., pp. iii–iv). It was in that tradition of scientific endeavor never divorced from the deepest philosophical spirituality that, I believe, Paolucci saw himself as squarely belonging in.}
7. Sappey 1847. {Cf. the fine historical section in this brilliant monograph, which gives a brief but thorough account of the history of the theories and observations concerning respiration in birds, ranging from Aristotle to the nineteenth century; ibid., pp. 62–81.}
8. Sappey 1847, p. 28, and Illustrations III and IV. {Cf. Figure 3.1.}
9. Sappey 1847, p. 21. He distinguishes between pulmonary diaphragm, which serves the dilatation of the lungs, and the abdominal-thoracic diaphragm, which intersects the cavity of the trunk.
10. Brehm 1872, p. 352.
11. Darwin 1871a, p. 343.
12. {Cf. Darwin 1871a, pp. 498, 634, for the studies by A. Newton and Murie. I have not found references for Cullen and Flower.}

Bibliography

Canestrini, Giovanni. 1869. *Compendio di zoologia ed anatomia comparata. Parte I: Vertebrati.* Milan: Gaetano Brigola.

Corti, Alfonso. 1851. "Recherches sur l'Organe de l'Ouïe". *Zeitschrift für Wissenschaftliche Zoologie* 3, pp. 109–169.

Galilei, Galileo. 1623. *Il Saggiatore.* Rome: Giacomo Mascardi.

Girardi, Michele. 1784. "Saggio di osservazioni anatomiche intorno agli organi della respirazione degli uccelli". *Memorie di matematica e fisica della Società italiana delle scienze*, II (1784), 2, pp. 732–748.

Helmholtz, Hermann von. 1863. *Die Lehre von den Tonempfindungen, als physiologische Grundlage für die Theorie der Musik*. Braunschweig: F. Vieweg und Sohn.

———. 1877. *Die Lehre von den Tonempfindungen, als physiologische Grundlage für die Theorie der Musik*. Vierte Aufgabe. Braunschweig: F. Vieweg und Sohn.

———. 1885. *On the sensations of tone as a physiological basis for the theory of music*. Translated by A. J. Ellis. 2nd English ed. London: Longmans & Co.

Menault, Ernest. 1868. *L'intelligence des animaux*. Paris: Libraie de L. Hachette et Cie.

Palmieri, Paolo. 2012. "Signals, cochlear mechanics and pragmatism: A new vista on human hearing?" *Journal of Experimental & Theoretical Artificial Intelligence*, 24, pp. 527–545.

Porro, Alessandro. 2001. "Girardi, Michele". *Dizionario Biografico degli Italiani*—Volume 56. Retrievable at: http://www.treccani.it/enciclopedia/michele-girardi_%28Dizionario-Biografico%29/.

Salvini, Anton Maria. 1735. *Discorsi accademici di Anton Maria Salvini, Gentiluomo Fiorentino Lettore di Lettere Greche nello Studio di Firenze e Accademico della Crusca: Sopra alcuni dubbj proposti nell'Accademia degli Apatisti*. Tomo secondo. Volume 2. Venice: Angelo Pasinelli.

Sappey, Ph.—C. 1847. *Recherches sur l'appareil respiratoire des oiseaux*. Paris: Germer Baillière, Libraie-Éditeur.

PART ONE, CHAPTER FOUR

The Fundamental Sounds of Birdsong

Casual observation of bird vocalizations might suggest that such sounds might be infinitely variable, and that they tend to escape conformity to law-likeness. Indeed, the variation in both power and pitch of the sounds is practically limitless. The former depends on the amount of force with which the air is pressurized against the thoracic larynx, and the latter on the number of vibrations made by the sonorous membranes during an interval of time. But this does not hold true in regard to the timbre, or color of the sound. I will argue that the economy of nature has not squandered its resources in the case of birds, any more than it did in the case of human speech. Although the number of words in human languages is almost limitless (and consequently the variability of acoustic articulations), the number of elementary sounds—the phonic roots, the raw materials of words, so to speak—is very small. In the final analysis, the raw materials of human vocalizations are vowels and consonants, which are combined in the formation of syllables in order to form any number of words, in the same manner in which quantitative values can be expressed with only a few digits. According to G. D. Whitney, twelve to fifteen *vocal attitudes* will suffice for any given language.[1]

First of all, I have ascertained that bird sounds can be either whistling or screaming. To a certain extent this classification corresponds to the different structure of the vocal apparatus. Indeed, the large majority of screaming and *tracheaphonous* birds have a screaming voice (recall the above discussion), while the majority of singing birds have a whistling voice. As far as this question is concerned, though, I regret that I was unable to extend my observations beyond the common European birds, whether domesticated or wild. Hence I have suggested this division according to my limited empirical knowledge. Screaming voices are more easily imitated by man because they are very similar to the sounds produced by mammals and by humans in particular. These screaming voices appear to be more common in birds that have less talent for singing, and indeed they are rare among singing birds. They sound harsh and dissonant, and therefore they are rarely employed by birds as sexual language. People refer to these voices as 'cries'. They perceive them as being similar to those of mammals. Occasionally such crying voices do indeed sound like faithful imitations of mammal voices. When the hoopoe (*Upupa epops*) arrives in our lands from the south, in May,[2] it produces the characteristic *hup hup*, which resembles the barking of dogs. This sound is the onomatopoeic basis of its name in many European languages[3] and Italian dialects.[4] In reference to the zoology of the island of Chiloe and of the Chonos archipelago, Darwin recalls that the bizarre *Guid-guid* (*Pteroptocus* or *Pteroptochos tarnii*), or *Black-throated Huet-huet*, belonging to the family of the *Rhinocryptidae*, imitates the barking of dogs so well as to deceive even the most inveterate observer.[5]

A greater prairie chicken observed by the traveler Douglas on the east coast of the North American continent made a rumbling tone, similar to a drum or the thunder, perhaps because it was convulsively expressing amorous passions, precisely as its European relatives do. The southern cassowary (*Casuarius galeatus*) produces two notes similar to a percussive sound when its intention is to draw attention to itself, but when it is angry it emits a sort of groan. The Arabs call the ruffed bustard (*Otis houbara*) 'Rhaad', which means thunder, because of the noise that the bird makes as it takes off. During the tranquil, moonlit nights in equatorial America it is common to hear a rolling tone in those immense *llanos*, particularly dull, which peters off melancholically, and which repeats itself for hours on end. Probably it is the love song of the gray-winged trumpeter (*Psophia crepitans*), in which anatomists discovered, close to the larynx, two resonant diverticula, shaped like hemispheres with chambers, in which the flow of air causes the sound by entering and exiting. The black crowned crane (*Grus pavonina*) of Papua New Guinea has a characteristic voice resembling a small trumpet. The peacock (*Pavo cristatus*) produces a low, guttural trumpet-like note before emitting its bright sonorous sound when it perches in the evening and at dawn. The harmonious *killkii* and the sweet *aug* (according to Naumann[6]) of the wild swan (*Cygnus musicus*) recall the sound of the violin, of the bell and the trumpet. The Nile goose (*Chenalopex Aegyptiacus*) trumpets around with little grace, making a harsh, dissonant, loud and incoherent sound (according to Brehm). The male of our great bittern (*Botaurus stellaris*) makes himself amorously conspicuous to the female with a bovine lowing. According to many naturalists, its sound would resemble that of cattle and could be heard at a distance. Neumann transcribed it with the word *üprumb* but had little luck in observing it. Vodzicki described, as reported in Brehm,[7] its mechanism as follows.

> I prudently got close and saw the female in shallow waters close to the male, with a swollen throat, poised like somebody at ease, perhaps like a Florentine music dilettante listening to a melody in a reverie. The ecstatic female, barely awake, stood in awe of the artist, a bass like Lablache.[8] The singer was standing on its feet, the body horizontal, the beak in water, and the mooing went on while the eater seemed like responding to the sound. After a few notes, I perceived the *ue* of Neumann, the male raised the head, moving it backwards, then after plunging the beak in water again, it made the mooing resound once more, at which point I was baffled. I realized that the sounds that initially resonate loudly are produced when the bird, after collecting water in its neck, pushes it with violent force. The music continued while he retired the head behind, so that I could no longer hear the notes….

I learned the song of the puffins (*Puffinus cinereus*) that live on the Adriatic coasts many years ago. It is mysteriously similar to the human voice. When the purity of the air and the tranquility of the sea allow the observer to pierce the immense expanses of blue and to listen to its silent solitude, it is easy to see the puffins gathered together at the water's edge in the clear light of the early mornings of May. They fly stylishly in various patterns, often abandoning themselves to the titillations of the breeze, cheerful, agreeing with one another, in a spirit of sincere friendship. It sounds as if they greet one another, laughing and enjoying nature. Their voices are sustained, low in register. They resemble the voices of sailors having funny conversations from their boats when whiling away the dead calm. But sometimes they repeat themselves, continuously and swiftly, like bursts of idle, crazy laughter. I had already done these observations when I read that the same facts had been noted by Darwin, who says of the innumerable flocks of puffins, seen by him in the island of Chiloe, that the noise they made resembled that of human beings that speak at a distance.[9] Finally, I will say something of the parrots, which, aside from learning to repeat sentences and words from humans by imitation, as permitted by their special anatomical apparatus, which is well suited for this type of voice, also cry out with more distinguished sounds. The sonorous *ka ka du*, or *kai kai du* earned the cockatoo its name, while the scarlet macaw (*Ara macao*), according to Burmeister,[10] would produce the harsh *ara*, or *arara*.

In addition to the above examples, the numerous series of species, whose voice resembles that of mammals and humans but which consists only of brief, broken, harsh accents that turn out to be of little interest to the casual observer, also belong in the group of the crying birds. These species are the majority of the entire family, far more numerous that the whistling species, as among the latter we only count most singing birds, some web-footed birds, and the waders. Moreover, whistling birds often produce some harsh sounds resembling those of crying birds, while it is rare that the latter make whistles.

Together with the crying and whistling voices one might count also other voices, more difficult to categorize, which I propose to call 'noisy' or 'a-rhythmical', in order to distinguish them from the former ones which are always musical. These noisy voices represent sounds without timbre, or color. They resemble, for instance, the collision of two solid bodies, the rustling of leaves, the fall of water, and so on. While, as we shall see, these voices are important from the viewpoint of psychic expression, they remain accessory, as they fail to tap the resources of the phonetic art. They may be used indifferently by both *tracheaphonous* birds and screamers, as well as by singers, though perhaps mostly by the latter. Let us now turn to the privileged group of those gentle, melodic creatures that can express their volitions and fascinate us with their harmonies, and which ever since remote antiquity have been admired by human beings.

In my view, the marvelous secret of their art consists in the extraordinary flexibility of their voice, capable of varying in force, pitch, and timbre, in sharp contrast to the crying birds mentioned above, whose voice always resembles, more or less, that of mammals and remains constant when it is emitted by the animal. In the long melodies of some singers, such as the nightingale, the robin, the lark, the goldfinch, the linnet, etc., one hears a tight succession of diverse accents that appears to be made by wind instruments, as though the animal passed instantly from the clarinet to the soprano clarinet, or from the flute to the piccolo. I have already discussed the causes of this sonorous ability by pointing out that they are to be found in the complex structure of the phonetic apparatus. I now propose to determine a genealogical order that systematically classifies the continuous series of voices produced by the singers in different groups of fundamental sounds.

Max Müller and other philologists put forward the claim that the first principles, or the elements of human languages are neither vowels nor consonants, although they form the raw materials of language, so to speak.[11] Indeed, they might be combined ad infinitum without producing a grammar or a dictionary. The elements of language consist of those fundamental sounds, *without any reality whatsoever in any naïve sense*, that are the raw materials of every language, and from which grammarians derived the series of alphabetical sounds, in the manner in which the bricks for building a house are fashioned out of clay. Now, if the diverse voices of singing birds, which are amongst the most talkative animals, truly constitute a language for their users, would it not be conceivable that fundamental sounds exist, i.e., typical timbres, from which the whole series of notes and songs are produced, exactly in the same way as, according to Müller, human languages were originally formed from elements that cannot be assumed to be real in any naïve sense? The question of the origin of birdsong does not seem, as far as I have been able to ascertain, to have been investigated. Ever since my childhood, I have had a special talent for imitating birdsong. Often I was so successful as to become a traitor. For, my talent served my passion for hunting birds. It was during my hunting trips that I made the vast majority of the observations. Hence my decision to write an essay on birdsong. By reflecting on the different modes in which I must articulate the lips, the tongue, the whistle, the cheeks, and the nose, in order to imitate birdsongs, I realized that a few fundamental sounds, constant in timbre, or color, but undetermined as to pitch and intensity, suffice to produce all the sounds made by these animals. These fundamental timbres would represent the natural instruments played by singing birds in their melodies.

In my research I could only observe our indigenous birds and some exotic ones. It is therefore logical to think that, if one could repeat my observations for bird genera in other continents—we know that Asia has 34, Africa 25, Oceania 43 and America more than 73—one would find different fundamental timbres, or instruments, than those commonly observed in Europe. On the other hand, it is true that Europe does not have a characteristic avifauna and no genera that do not have their exotic counterparts. Thus, I am confident that my observations are representative enough.

All singing birds can compose more or less harmonious stanzas, occasionally very complicated, starting from a few fundamental timbres of different color that I will describe as best as I can.

The first, most essential and perhaps most common timbre is a species of musical color that I will call 'supple, or mild timbre', resembling a sound between that of the piccolo and the soprano clarinet. It is very sweet to the ear, especially in lower and more sustained notes. It is most frequently used by singing birds. It is the basic sound of the song of the canary and of most woodland birds. It is possible to imitate it with great fidelity by barely opening the mouth, bringing the lateral borders of the tongue against the superior teeth arches, and by letting the air flow out from a narrow escape formed by the palatal vault and the pursed tip of the tongue. Indeed, it is easy to imitate it, and with practice even the most delicate modulations can be reproduced by aptly articulating the lips. The more tight and protruding the lips, the lower the sound. It becomes gradually higher when the lips take a laughing attitude. As for the intensity of the sound, the mouth can only reach a certain value beyond which it becomes impossible to reproduce the original faithfully. A derivative of the supple timbre might be a flute-like sound, which can be rarely heard among our birds. It might thus be called 'flute-like timbre'. The bullfinch, the oriole, and the cuckoo, for instance, can make it.

The second fundamental timbre is that which is sometimes heard most clearly from certain birds, and which is popularly called 'mute', or *sordino* in Italian, in a special sense on which I will say more later on. It resembles the supple timbre of which it might possibly be considered a most acute variation. It can be easily imitated by most people, by protruding the lips forwards and making them match all along the internal border, except in the middle, where a small opening must be left open. I will call it 'acute timbre'. It is a sibilant whistle which can be rarely heard in birdsongs. It is more often used as a warning signal.

The third fundamental timbre might be called 'harsh' in relation to the nature of the voice from which it can be recognized. Many birds make use of it producing an infinite variety of modulations, both in standalone patterns and as an interpolation between the stanzas of sexual songs. It can be heard in the calling song of the common sparrow, in the voice of the *Sylvia hypolais*, or *Sterpazzola* in Italian, in the song of the *Lanius* passerines, in the call of the mountain sparrow (*Petronia stulta*), of the mountain finch (*Fringilla montifringilla*), in the whistle of fear of the crested lark (*Alauda cristata*), in the call of the European greenfinch (*Loxia chloris*), and in other birds. It can be easily imitated by placing the tongue in the same way as in the supple timbre while slightly making the superior teeth arches touch the inferior lip and flexing the buccinators.

The fourth fundamental timbre is the most difficult to imitate. I have yet to find a bird-seller who can reproduce it faithfully. It might be called also 'trilling' timbre, as it is a sort of whistle, blunt, more often acute than low, like a tremolo accompanied by a weak roar. It is used alone in certain patterns by a few birds, such as the common sparrow, the great tit (*Parus major*), the common swift (*Cypselus apus*). It is part of the voice of some birds such as the common linnet (*Linota cannabina*). At any rate it is a timbre of great importance, as we shall see better when we discuss the expressivity of song in different families of singing birds. To sum up, the supple timbre, the acute timbre, the harsh timbre, and the trilling timbre represent the fundamental sounds which, when they are amalgamated together, modulated and patterned according to musical rhythmic structures, can reproduce and imitate the voices of all European birds.

I acknowledge that it would be very difficult to imitate the voices of birds for those who have not become accustomed to them by dint of practice and observation. Hunters and bird-sellers make use of simple wind instruments and whistles, by which they can imitate the song of a few birds such as the lark, the thrush, the blackbird, and a few others. But these instruments can only approximate the sounds based on the supple timbre. All others are out of the question.

The syllables that have been proposed so far in the ornithological literature as principle means of transcription of birdsongs may be accurate enough for those screaming voices that, as already noted, resemble the human voice, but they are useless when it comes to whistling birds, where the articulation of consonants is missing, and where instead what is most characteristic is the timbre of the sound. On this broader question, the works by Brehm,[12] Bechstein,[13] and Naumann,[14] remain fundamental, in addition to important contributions by Wilson, Rosenberg, Lesson, Gerardt, Waterton, Poepping, Schomburg, Audubon, Lenz, Savi, Salvadori and others.[15]

Notes

1. Whitney 1876. {The Italian translator of the book, Francesco D'Ovidio, noted in the Preface that the subject matter of this study by Whitney, a Sanskrit scholar, is first and foremost the philosophy of language. "This book is above all devoted to the philosophy of language, that is, it emphasizes the nature of language, the relation of language to thought, language learning in humans, and investigates the origin of language" (ibid., p. xi). D'Ovidio goes on to expound his fascinating views on the relationship between science and philosophy and criticizes the arrogant role that, he argues, philosophers often play in superficially appropriating the discoveries made by scientists. On the other hand, he points out that the scientific method, which, in his view, demarcates the so-called positive sciences from philosophy, cannot and should not be divorced from philosophical speculation, and that the latter is indeed paramount in forging worldviews that create the milieu in which science can thrive. His most original thesis regards the 'specialist philosopher of science'. It is this ideal figure who, according to D'Ovidio, has both the scientific competence and the vision to transform the positive data of science into the worldview to which the human intellect must ultimately aspire. I think that D'Ovidio arguments are relevant for our understanding of the type of scientific-philosophical research that Paolucci intended to carry out. His approach to the study of birdsong is definitely rooted in solid observational zoology as well as in comparative anatomy and physiology, as we have seen so far, but he never loses sight of the fundamental importance of philosophical speculation. He is both the scientist and the specialist philosopher of science described by D'Ovidio.}
2. Paolucci 1873. {I have corrected the title of the paper referred to by Paolucci (see bibliography), as the one cited by him in this footnote appears to be slightly different that the one printed in 1873. I have also added the page numbers, not reported by Paolucci.}
3. Our *Bubola* is the *Upupa* of the ancient Latins, epops in Ancient Greek, mpoufos in modern Greek {here Paolucci used the Greek alphabet, which I could not reproduce}, *Huppe* in French, and *Abubilla* in Spanish.
4. Pokorny 1876, p. 79. {Here Paolucci also cites an *Ornitologia Italian*a, at p. 46, by Tommaso Salvadori, presumably published in Milan in 1874. The citation, I believe, might be an error. I have not been able to find this work by the ornithologist Salvadori. On the other hand, an *Ornitologia Italiana* in three volumes, by Paolo Savi, was published posthumously in 1873–1876, but in no volume does the page referenced by Paolucci seem to be relevant to his discussion.}
5. Darwin 1872, p. 251.

6. {Probably the ornithologist Johann Friedrich Naumann (1780–1857).}
7. Brehm 1870, p. 741.
8. {Luigi Lablache, Italian singer, famous for his roles, like Leporello in Mozart's *Don Giovanni.*}
9. Darwin 1872, pp. 252 ff.
10. {Probably the naturalist Hermann Burmeister (1807–1892).}
11. Müller 1870–1871, II, p. 83. { In essence, Müller offered a cogently argued, organicist view of language, together with an ontology of thought and language as being one and the same living entity, and last but not least an analytical approach to linguistics that must have appealed to Paolucci. Müller stated that "[g]ranting, however, to the fullest extent, the one and indivisible character of language and thought, [...] we may yet, I think, for scientific purposes, claim the same liberty which is claimed in so many sciences, namely, the liberty of treating separately what in the nature of things cannot be separated. Though color cannot be separated from some ethereal substance, yet the science of optics treats of light and color as if they existed by themselves. [...] It is the same in language, and though I consider the identity of language and reason as one of the fundamental principles of our science, I think it will be most useful to begin by dissecting the dead body of language, by anatomising its phonetic structure, without any reference to its function, and then to proceed to a consideration of language in the fullness of life, and to watch its energies, both in what we call its growth and its decay. I tried to show in my first course of lectures, that if we analyze language, that is to say, if we trace words back to their most primitive elements, we arrive, not at letters, but at roots. This is a point which has not been sufficiently considered, and it may almost be taken as the general opinion, that the elements of language are vowels and consonants, but not roots. If, however, we call elements those primitive substances the combination of which is sufficient to account for things as they really are, it is clear that we cannot well call the letters the elements of language; for we might shake the letters together ad infinitum, without ever producing a dictionary, much less a grammar". Cf. Müller 1873, II, pp. 79–80. However, I believe that Paolucci's work, both in its original intent and in its substantial achievements, emphasizing that birdsong is a language, felicitously undermines Müller's contention that a radical, ontological separation would exist between human and non-human animals, in so far as, in his opinion, the latter do not have 'reason'. Indeed, Müller argues, "I shall not be understood as here denying the reality of thought or mental activity in animals. Animals and infants who are without language, are alike without reason; but the difference between animal and infant is, that the infant possesses the healthy germs of speech and reason, only not yet developed into actual speech and actual reason, whereas the animal has no such germs or faculties, capable of development in its present state of existence. We must concede to animals 'sensation, perception, memory, will and judgment,' but we cannot allow to them a trace of what the Greek called *logos*, i.e. reason, literally, gathering, a word which most rightly and naturally expresses in Greek both speech and reason" (ibid., p. 66).}
12. Brehm 1870, 1872.
13. {I have not been able to pinpoint and consult the 1791–1793 original, German edition of Bechstein's work on German birds cited by Paolucci, but a subsequent edition is Bechstein 1801–1809, which carries a title similar to that cited by Paolucci as an added title

to volumes two, three, four. Cf. also the French edition, Bechstein 1838 (Paolucci cites an 1828 edition, presumably the first, that I have not been able to consult).}

14. Naumann 1820–1860.
15. I believe I have identified the following authors: Alexander Wilson (1766–1813), Hermann von Rosenberg (1817–1888), René Primevère Lesson (1794–1849), Charles Waterton (1782–1865), Eduard Friedrich Poeppig (1798–1868), John James Audubon (1785–1851), Tommaso Salvadori (1835–1923), and Paolo Savi (1798–1871), Harald Othmar Lenz (1798–1870).

Bibliography

Bechstein, Johann Matthäus. 1801–1809. *Gemeinnützige Naturgeschichte Deutschlands nach allen drey Reichen: ein Handbuch zur deutlichern und vollständigern Selbstbelehrung besonders für Forstmänner, Jugendlehrer und Oekonomen.* Leipzig: Siegfried Lebrecht Crusius.

———. 1838. *Manuel de l'amateur des oiseaux de volière.* Déuxiéme Édition. Brussels: Société Belge de Librarie.

Brehm, Alfred. 1870. *La vita degli animali, descrizione generale del regno animale. Traduzione Italiana dei Professori G. Branca, S. Travella, e altri naturalisti. Riveduta da Michele Lessona e Tommaso Salvadori. Volume Quarto.* Turin: Unione Tipografico Editrice Torinese.

———. 1872. *La vita degli animali, descrizione generale del regno animale. Traduzione Italiana dei Professori G. Branca, S. Travella, e altri naturalisti. Riveduta da Michele Lessona e Tommaso Salvadori. Volume Quinto.* Turin: Unione Tipografico Editrice Torinese.

Darwin, Charles. 1872. *Viaggio di una naturalista intorno al mondo.* First authorized Italian translation by Michele Lessona. Turin: Società L'Unione Tipografico Editrice.

Müller, Max. 1870–1871. *Nuove letture sopra la scienza del linguaggio.* Edited by Gherardo Nerucci. Volume Primo e Secondo. Milan: Treves.

———. 1873. *Lectures on the science of language.* Seventh edition. 2 vols. London: Longmans, Green, and Co.

Naumann, Johann Andreas. 1820–1860. *Naturgeschichte der Vögel Deutschlands, nach einigen Erfahrungen entworfen: Durchaus umgearbeit, systematisch geordnet, sehr vermehrt, vervollständigt, und mit getreu nach der Natur eigenhändig gezeichneten und gestochenen Abbildungen aller deutschen Vögel, nebst ihren Hauptverschiedenheiten.* Leipzig: G. Fleischer.

Paolucci, Luigi. 1873. "Gli uccelli migratori nella provincia di Ancona. Primi studi." *Atti della Società Italiana di Scienze Naturali* 16, pp. 113–137.

Pokorny, Alois. 1876. *Storia illustrata dei tre regni della natura. Parte prima. Regno animale.* Edited by Michele Lessona and Tommaso Salvadori. Second edition. Rome, Turin, Florence: Loescher.

Savi, Paolo. 1827–1831. *Ornitologia toscana; ossia, Descrizione e storia degli uccelli che trovansi nella Toscana.* Three volumes. Pisa: Dalla Tipografia Nistri.

Whitney, William D. 1876. *La vita e lo sviluppo del linguaggio.* Translated with anotations by Francesco D'Ovidio. Milan: Fratelli Dumolard.

PART ONE, CHAPTER FIVE

Musical Representation of Birdsong

In order that the expression and the imitation of birdsong be plausible and consistent, we must take into account the inherent qualities of sound, namely, intensity, pitch, and timbre. It is clear that by using syllables it will be impossible to describe any of these qualities. Notwithstanding all the complexity of their doublings, triplets, and of their grave, acute, and circumflex accents, syllables will at most be able to mimic sound vaguely, and they will only serve as guidance and as aiding schematics for those who already know the birdsongs. Furthermore, syllabic interpretation is different, according to the written and spoken language of the observer, and cannot but be coupled with the articulation of consonants that is totally absent from melodic voices. When a singing bird expresses a long stanza, or perhaps a measured succession of accents, there is always present what is known as musical rhythm, or movement, that is, the measure of time reckoned in bars, or parts of a bar, as I have been informed by illustrious music experts.[1] The simple articulation of sound in syllables, that is, the sounds of syllabic language, cannot achieve the imitation of birdsong. Music notation is in this regard more useful. Indeed, in my view, it is the only means of notation successfully applicable to birdsong. I will offer an example to clarify matters further.

Of the nineteen sounds that he studied in the finches of Schnepfenthal Lenz reports one that he calls 'third dampener' as follows.[2] "Disdisdistritritriclapclapclapzihe". As far as I have been able to understand it, I speculate that this syllabic pattern must be the very song that a finch, which I caught at the beginning of autumn and domesticated, has repeated to me for years. The readers should ask themselves if they understand the above transcription, or form an idea of the song based on it. On the other hand, I tried a musical translation, as follows.

FIGURE 5.1 Paolucci's music notation of the finch's song (Paoucci 1878, p. 48). Caption by Paolo Palmieri.

I believe that the music notation I have suggested captures faithfully the tempo of the finch's song.

Thanks to music we can also notate the two other salient qualities of sound, i.e., intensity and pitch, as these are the same that are commonly used to express the melodic lines of songs. The greatest difficulty, however, lies in rendering the color or timbre of the sound. For, in this case, music itself as it is usually notated is of little help in that no music instrument can imitate the whistling voice of a bird, let alone its song, especially when it comprises more than one timbre. On the other hand, imitation must be perfect because the smallest variation will change the nature of the song and indeed distort its meaning. I could ascertain the latter fact by observing that birds do not answer nor listen to an artificial call if it differs from their own even only in the most delicate nuances. Having described as best as I could the nature of the fundamental timbres, I have added a few conventional letters to the notes in the examples in order to facilitate the comprehension of the transcriptions. Thus, 'ac' will indicate the acute timbre, 'as' the harsh timbre, 'tr' the trilling timbre, and no letter will indicate by default the supple timbre. The accents of some whistling birds, and often of singing birds, come roughly close to a syllabic pattern. In the case of birds that are rather common, people have named them according to onomatopoeia, distinguishing different species. I will furnish a short catalog and report the synonyms that are common in different regions of Italy, while also translating the syllabic forms into music notation, so as to illustrate the method I have described.[3]

FIGURE 5.2 Examples 1–11 (from above) as reported by Paolucci (Paolucci 1878, pp. 50–53). Caption by Paolo Palmieri.

FIGURE 5.3 Examples 12–20 (from above) as reported by Paolucci (Paolucci 1878, pp. 54–56). Caption by Paolo Palmieri.

1. *Athene noctua* (Italian owl). Onomatopoeic synonyms. Cucumeggia (from Lucca), cucumiao, cuccumeu (Sardinia). Syllabic expression: ko ko bio—ko ko miao.[4]
2. *Ephialtes scops* (Italian *assiolo*). Onomatopoeic synonyms. Chiocch (Pidemont), chiú (Lombardy), chiò (Venice), ciò (Parma), cios (Piacenza), ciù (Modena and Romagna), chiù (Florence, Siena, Rome, Marghe), chiù (Sicily). Syllabic expression: chiù, chiù.

3. *Cuculus canorus* (Italian cockoo). Onomatopoeic synonyms. Cuccuc (Venice), coucoù (Pidemont), Cucco (Genova, Tuscany, Marche, Rome, Naples), cucu (Sardinia), ku-ku (Malta). Syllabic expression: cu-cuc.
4. *Merops apiaster*. (Italian *gruccione*). Onomatopoeic synonyms. Grùcciolo, gràvolo (Rome), barbaraschio (Fermo), guaraguasto (Ancona), grallo, gaulo, goro (elsewhere). Syllabic expression: a) scirr scirr, b) croù, croù.
5. *Cypselus apus* (Italian *rondone*). Onomatopoeic synonyms. Pivi (Pidemont), bibì (Vercelli), spirli (Corsica). Syllabic expression: a) spi spi—cri cri, b) briiih, briiih.
6. *Parus coeruleus* (Italian *cinciarella*). Onomatopoeic synonyms. Cincia (Pidemont), ciribibi (Saluzzo), bibi-zal (Modena), zerepella (Naples). Syllabic expression: a) zi te re te te, b) ti ciò ciò cià cià.
7. *Parus major* (Italian *cinciallegra*). Onomatopoeic synonyms. Cibibì, ciribibì, ciribin, Liarì (Pidemont), tardaci (Fermo), cincipottola (Florence), cirincio, cirlincio (Sicily). Syllabic expression: cîs pin pin.
8. *Phillopneuste rufa* (Italian *Luì piccolo*). Onomatopoeic synonyms. Luì (Tuscany), ciaucin, tuich (Pidemont), tuì, tuin (Lombardy), tuit, buit (Cuneo), ciuin (Venice), puì (Rome), ciuì (Ancona). Syllabic expression: chiuì, chiuì.
9. *Montacilla alba* (Italian *ballerina*). Onomatopoeic synonyms. Spasaclì (Brescia), cutréttola (Florence), pispisa (Sicily), zic-zac (Malta). Syllabic expression: vitiè—vitiè.
10. *Budites flavus* (Italian *cutrettola gialla*). Onomatopoeic synonyms. Cutti (Pisa), guzzì, cuzzì (Rome). Syllabic expression: vi zi.
11. *Anthus pratensis* (Italian *pispola*). Onomatopoeic synonyms. Ciè-ciè (Como), spiplèn (Modena), sì- sì (Genova), babusso (Rome, Ancona), si si (Macerata, Ancona). Syllabic expression: a) isst isst—isst isst isst, b) sì sì—sì sì sì.
12. *Alauda arborea* (Italian *tottavilla*). Onomatopoeic synonyms. Lodovigh, sarlodì (Lombardy), cutuì (Modena), tuttuilla (Rome), cuclaita (Malta). Syllabic expression: toto-vì toto-vì—totigliè.
13. *Miliaria Europaea* (Italian *strillozzo*). Onomatopoeic synonyms. Ciattaron (Genova), stiattajone (Pisa), striglio (Marche), ciciruni (Sicily). Syllabic expression: zi zi zi, zizizisrrrrrr.
14. *Emberiza cirlus* (Italian *zigolo*). Onomatopoeic synonyms. Zizì (Pidemont), sia (Milan), zivola (Rome), zigola (Ancona), Zita (Macerata), zinzicula (Catania), zinzia (Sardinia). Syllabic expression: zi zi.
15. *Emberiza cia.* (Italian zigolo mucciatto). Onomatopoeic synonyms. Zia, sia (Lombardy). Syllabic expression: ziiie ziiie.

16. *Fringilla coelebs* (Italian *fringuello*). Onomatopoeic synonyms. Schinsoun (Pidemont), quinquìn (Susa), pincione, spincione (Rome), pinzumi (Sicily), spunzun (Malta). Syllabic expression: finch finch.
17. *Oriolus galbula* (Italian *rigogolo*). Onomatopoeic synonyms. Ouriéul, ouriéu (Pidemont), rigogolo (Florence), gràvolo (Rome), graulo (Ancona), gajulu (Sicily). Syllabic expression: di—tleò, gidadi—tleò.
18. *Chrysomitris spinus* (Italian *lucarino*). Onomatopoeic synonyms. Lugarin, tarin (Pidemont), legorin, lugarì (Lombardy), ogarì (Ancona). Syllabic expression: luglî, luglî.
19. *Aegialites euronicus* (Italian *corriere piccolo*). Onomatopoeic synonyms. Cirrivì (Messina), ciurlì (Ancona). Syllabic expression: hidìdì.
20. *Totanus fuscus* (Italian *chiò-chiò*). Onomatopoeic synonyms. Chiò-chiò (Pisa), ciuvet (Malta). Syllabic expression: chiù-chiù.

Also screaming birds occasionally produce sounds that have been noted by people and chosen to form onomatopoeic names. In such cases the similarity was even more striking than in the case of whistling birds as the voice of screaming birds can easily be imitated with great fidelity thanks to the human vocal apparatus. A good example is the raven-lord (*Corvus corax*) that in some Italian dialects is known as *corv*, *crov*, *crova*, *crobu*, in French *corbeau*, in German *Rabe*, while the common call of this bird can easily be expressed with the harsh syllables *kork*, *kork*, or *rabb*, *rabb* (according to Brehm), from which one can compose the well-known sound pattern of the ravens during the mating season. This sound must have inspired Dupont de Nemours's weird fantasy in his *Dictionaire des corbeaux*, a work that cost him, according to the author, 'deux hivers, et grand froid aux pieds et aux mains'.[5]

FIGURE 5.4 The raven picturing the stage of *nigredo* (blackness) within the alchemical vessel. Paolucci's quotation of Du Pont De Nemours's work on ravens brings up the image of a 'sauvage réduit', a safe haven in the wilderness, in which Du Pont De Nemours, as he tells us, receives illumination from the ravens that so fascinate him. The enclosed space of the 'sauvage réduit' suggests the symbolism of the alchemical vessel. From: Andreas Libavius. *Alchymia Andreae Libavii, recognita, emendata, et aucta; tum dogmatibus & experimentis nonnullis.* Francofurti : excudebat Ioannes Saurius, impensis Petri Kopffii, 1606, p. 701. Retrieved at http://www.e-rara.ch/cgj/alch/content/titleinfo/1912930: Stiftung der Werke von C.G.Jung, Zürich, online, http://dx.doi.org/10.3931/e-rara-7368 / Public Domain Mark. Caption and figure by Paolo Palmieri.

Especially in order to transcribe musically the voice that in many species of birds has originated the onomatopoeia, I needed to have recourse to indigenous species, whose songs I knew exactly. However, by availing ourselves of the syllabic notation we will discover that many species of exotic birds have names that remind us, albeit not perfectly, of the peculiar form of their phonetic language. To mention a few examples, we have the great kiskadee (*Saurophagus sulphuratus*) of Malndonado, the pied plover (*Vanellus cayanus*), the black-throated huet-huet (*Pteroptochus tarnii*), the *Cacatuidi*, the northern screamer (*Chauna chavaria*), and others.[6]

To summarize what we have said in regard to the sonorous elements of birdsong, that is, to its musical patterns, we will conclude by pointing out that it differs from the vocalizations of inferior animals (such as insects), in that it is always modifiable by the birds at will. It is produced by a pneumatic apparatus, whose essential parts are the mobile tracheal duct, the inferior larynx, and air pockets. Although birdsong is infinitely variable, a few sounds or fundamental timbres, which we called the supple, the acute, the harsh and the trilling, are sufficient to form it and express its most delicate nuances. Finally, we noticed that music is often far better suited to transcribe it than the syllabic notation. In the remainder of this work, we will investigate the biological functions of the canorous attitudes whose underlying anatomical structures and physiology appear so perfectly contrived.

Notes

1. I would like to thank master G. Grassoni warmly, for the thorough advice he gave me in matters concerning the musical interpretation of birdsong in the course of my studies.
2. {Cf. Lenz 1851, pp. 152 ff., where Lenz talks about the song of finches. However, I have not been able to identify the syllabic pattern quoted by Paolucci. He may have consulted a previous edition of this work, or a different work altogether. He does not cite the work and the page from which he quotes the transcription, and the title page of the third edition I could peruse announces that it is the 'dritte sehr veränderte Ausgabe'.}
3. On the synonyms, cf. Pokorny 1876, Savi 1827–1831, III. {Here again Paolucci cites an *Ornitologia Italiana*, by Tommaso Salvadori, presumably published in Milan in 1873. I have not been able to find this work by the ornithologist Salvadori.}
4. {In the catalog of twenty examples given by Paolucci, I have omitted the abbreviations presumably referring to names of the ornithologists according to whose nomenclatures Paolucci reports the syllabic expressions of the voices, or their onomatopoeic names. Their identification is uncertain as I have been unable to disambiguate the short abbreviations (the only clear case seems to be Alfred Brehm, from whom most syllabic expressions are reported by Paolucci without reference to the specific work). The curious reader should consult Paolucci's original work.}
5. Du Pont De Nemours 1813. {Paolucci's reference to Du Pont De Nemours invites speculation. Du Pont De Nemours's natural histories evoke (whether consciously or not, I do not know) an undercurrent, flowing throughout modern European history, of a naturalistic style of being-in-the-world that eschews the remorse-less and indiscriminate killing of animals for allegedly 'innocent' research purposes, rejects the scholarly study of death and the necrophilia of the pleasures of the dead bodies of animals, and instead affirms the

sacred union of the human life-world with the animal life-world. Du Pont De Nemours says of the ravens, in his unprepossessing but poetic French,

> [j]e voudrais vivre avec eux aux champes, m'y éclairer de leurs lumières, et les mener aussi quelquefois loin du village dans un sauvage réduit, bien immobiles, bien silencieux, l'oeil au guet, l'oreille attentive, un crayon et un petit livre blanc à la main, les corbeaux, ni les autres animaux n'ont pas peur des livres. Là, j'inviterais mes dignes amis à étudier la *nature vivante*, et à noter leurs remarques sous son auguste et correcte dictée. Ils apprendraient beaucoup de mots du dictionaire de plusiers espèces. C'est un travail long. Les corbeaux m'ont coûté deux hivers, et grand froid aux pieds et aux mains. (ibid., p. 177)

No wonder that Paolucci feels the mystical power of Du Pont De Nemours' 'strana fantasia' (weird fantasy). It is beyond the scope of this commentary to delve into the deeper linguistic strata of the lexicon employed by Paolucci, but the adjective 'strano' cannot be mistaken for a simplistic comment on Du Pont De Nemours 's imagination. To every educated Italian it will awaken the memories of its formidable presence in Dante's *Commedy*; for instance, in the context of diabolical magic and uncanny inhumanness, or better, beyond-humanness. Its resurfacing in Paolucci's text, and its emerging in his imagination at this juncture, in close proximity with Du Pont De Nemours's image of the 'sauvage réduit', a metaphor for the alchemical vessel (let us recall the image of the raven, as picturing the stage of *nigredo*. in alchemical symbolism), conjures the possibility of the trans-species fusion of life-worlds that Paolucci's assimilation of birdsong to music must entail.}

6. {I have been unable to identify a third family in this list that Paolucci calls 'Arare'.}

Bibliography

Du Pont De Nemours, Pierre Samuel. 1813. *Quelques mémoires sur différens sujets: la pluspart d'histoire naturelle, ou de physique générale et particulière*. Seconde Édition. Paris: De l'imprimerie de A. Belin.

Lenz, Harald. 1851. *Gemeinnützige Naturgeschichte: Vögel, Band 2*. Dritte Ausgabe. Gotha: Beckersche Buchandlung.

Paolucci, Luigi. 1878. *Il canto degli uccelli. Note di fisiologia e biologia zoologica in rapporto alla scelta sessuale e alla lotta per l'esistenza*. Milan: G. Bernardoni. Retrievable online at: http://dfgviewer.de/show/?set[mets]=http%3A%2F%2Fwww.zvdd.de%2Fdms%2Fmetsresolver%2F%3FPPN%3DPPN641658559&set[image]=1

Pokorny, Alois. 1876. *Storia illustrata dei tre regni della natura. Parte prima. Regno animale*. Edited by Michele Lessona and Tommaso Salvadori. Second edition. Rome, Turin, Florence: Loescher.

Savi, Paolo. 1827–1831. *Ornitologia toscana; ossia, Descrizione e storia degli uccelli che trovansi nella Toscana*. Three volumes. Pisa: Dalla Tipografia Nistri.

PART TWO

Patterns of Psychic Expression in Birdsong

PART TWO, CHAPTER ONE

Intelligence and Communication in Birdsong

It is all too easy to invent a science arbitrarily whenever observation is difficult or experiment impossible. This is confirmed by those who venture to speculate. They start from dogmatic convictions that are not grounded in facts and sound empiricism. The questions of animal psychology, which I will presently raise, might appear to be 'speculative' in the pejorative sense I indicated, especially if they were considered from the interpretive perspectives that have been popularized by too many visionaries. Sundry writers and naturalists have taken up the discussion of animal intelligence since antiquity and have touched on the question of animal vocalization and communication. Opinions are numerous and widely different. Some exaggerated the intellective capacities of animals to the extent of attributing to them both reason and thought. Others took the opposite view and believed that even the most perfect animals are material automata, whose vital phenomena are but the result of external nature acting on their senses. Dupont de Nemours interpreted birdsong as an expression of animal cognition in a very peculiar sense. Aside from the dictionary of ravens and other strange things, he gave us a translation of the nightingale's love song into French verses.[1]

> Dors, dors, dors, dors, dors, dors, ma douce amie,
> Amie, Amie
> Si belle et si chérie!

Dors en aimant,
Dors en couvant,
Ma belle amie,
Nos jolis enfants;
Nos jolis jolis jolis jolis jolis
Petits enfants
(A pause)
Mon amie
Ma belle amie,
A l'amour,
A l'amour ils doivent la vie,
A tes soins il devront le jour.
Dors, dors, dors, dors, dors, dors, ma douce amie,
Auprès de toi veille l'amour,
L'amour,
Auprès de toi veille l'amour.

This is poetry echoing certain ancient philosophical systems on the nature of animals and their souls, especially the skeptical philosophies. Plato and Pythagoras thought that animals have a rational soul. Aristotle found in them the hallmark of genius.[2] Plutarch recognized reason in animals as the principle of their industriousness.[3] Seneca thought that their capacities amounted to infused science (Epistle 121).[4] A few moderns followed in the footsteps of this ancient spiritualist school of thought. Montaigne is an example. He conceded that animals have language and concluded that they might well regard us as beasts, as much as we regard them as such, for the simple reason that beasts and humans do not understand one another.[5] In a valuable collection of animal stories and lore, E. Menault exalts the psychic powers of animals. His book is animated by a poetic imagination throughout, and he goes so far as to translate the love song of a bird as follows. "Mais aimez-vous, aimez-vous, aimez-vous donc. Il faut vous aimer, toujours vous aimer, rien que vous aimer!"[6]

In contrast to these opinions about animal intelligence, other authors deny that animals have any of the intellectual traits that, according to them, distinguish humans from non-human animals. Galen opposes that view that animals have the possibility of accomplishing any vital function with intelligence. Descartes claims that animal patterns of behavior can be explained according to mechanism, as necessary consequences of the influences of external agents. Buffon is more or less in agreement with Descartes, although he acknowledges that animal life is endowed with the capacity for the sensations of pain and pleasure. E. Reimar has treated the subject of animal intelligence on the basis of numerous scientific and every-day life observations. He almost always finds an analogy between the psychic acts of animals and humans. He would also concede that animals have free choice. However, he opposes the identity of nature between human beings and non-human animals.[7] I might add many other names of partisans of one or the other viewpoints, but I will refrain from doing so for the sake of brevity.

These two philosophical schools are extreme. The first equates non-human animals with humans, the second turns them into automata. But there has always been a savvier middle ground. For centuries an intermediate view has been discussed according to which the difference between human and non-human animal cognition does not consist in the nature of cognitive powers but rather in their degree. In the fourth century AD, the Bishop Nemesius anticipated the rationalistic school. He wrote: "It seems that the Creator has subtly connected the objects of different nature to one another, so that thanks to this link they could form a whole". Furthermore, he said, "…as He desired to pass from irrational animals to rational human beings, He accomplished this by degree, endowing the animals with natural lights, in order that they could approximately reach to the rational creatures".[8] The majority of modern naturalists, including the proponents of Darwinian theories, among whom Lamarck was the initiator, agree on the concept of gradation in describing the relation of animal and human cognition. A. De Quatrefages is very clear in this regard, although he remains doubtful as to whether species are fixed or subject to evolution. "The animal has its share of intelligence, and its fundamental faculties are not different, in the last analysis, although they are less developed than ours. The animal senses, wills, remembers, reasons, and the exactitude and certainty of its judgments are often extraordinary, while the errors that we see them make all the time demonstrate that these judgments are not the result of a blind and determined force".[9] In a fine study of comparative psychology, Tito Vigoli reached the same conclusions as Quatrefages's. Indeed, Vignoli goes further than those who share his views on animal cognition. He strives to demonstrate that instinct is inherited like a habit that originated in intelligence.[10] As Condillac had said much earlier: "Instinct is but habit deprived of reflection".[11]

According to modern evolutionary theorists, the nature and causes of intelligence in animals are identical with the nature and causes of intelligence in humans. The difference consists only in the degree of intelligence, although such difference is certainly enormous. Animals realize the powers of the brain as far as this is made possible by its smaller or greater perfection. As far as my subject is concerned I agree with them. Birdsong is a rudimentary form of language. This is what I will, hopefully, be able to demonstrate in the remainder of this work. So far I have expounded the anatomical basis of the singing apparatus. Insofar as language, then, birdsong is a powerful means in the struggle for survival of the individual and in the sexual competition that aims at the improvement of the species, which changes by continually taking on novel characters and losing those that have become useless owing to their not being used any longer.

Although they are inferior to the mammals in terms of organization, birds often manifest great expenditure of energy during their entire life. This is observed in both their vegetative functions and their relational ones. In certain respects they are superior to all warm blooded vertebrates. The frenzy of their existence rivals that of butterflies and mayflies. Their body temperature, which is higher than that of humans by almost five degrees, maintains them in a heightened state of hyperesthesia and make them capable of enduring hardship. They do not hibernate. Both their young and old ages are short in comparison with youth and maturity. Most of them do not sleep long. I observed that the softest noise wakes them up. The propensity for flight of the *laridae* (gulls), of the *procellariidae* (sea birds) and of the sparrows is proverbial. Since the time of Lazzaro Spallanzani it was known that they emigrate to Senegal. It is believed that the common swifts (*Cypselus apus*) can spend their summer nights in the high regions of the sky,[12] and I heard their voices while walking along a beach in Ancona during a quiet summer night.

As to the phenomena of cognition in birds innumerable facts have been observed that suggest that they should be classed with mammals in this respect. Who does not know, for instance, the memory of the magpie, the cleverness of raven and the swallow, the circumspection of the starling, or the tenderness of the dove? Who has not read the moving stories of the famous Iako (*Psittacus erythacus*) who died of pain for the loss of its master Kleinmayrn?[13] Or the anecdotes of the parrot that replaced Dutch words with German ones when the latter were missing in certain phrases?[14] And what about the parrot owned by Wood, who adopted abandoned birds and used to carry them proudly on his shoulders?[15] Many interesting accounts have been given by Schinz[16] and by Cornely de Saint Gerlach on the intelligence and attitudes of the long-billed chough (*Fregilus graculus*).[17] One day I was stealing the babies from the nest of a couple of larks (*Galerida cristata*) when the mother arrived with her mouthful of food. All at once, impelled by her great love for the little ones, she plucked up her courage to get bravely close to me. When she realized my cruelty, she was overwhelmed by pain and desolation and was seized by convulsions in a turmoil of screams.[18] These examples should suffice to show that nature generously bestowed on birds both physical prowess and moral capacities. Once we have acknowledged the high degree of their intelligence and the exquisite sensitivities with which they partake of the noblest sentiments of life, we cannot but hold them in the highest esteem when it comes to considering the most admirable of their talents, namely, song. Shall we say with Flourens, that "their voices [i.e., those of animals in general], their vocalizations, their accents, are but the *forced and unwilled* expression of sentiments?, or that their voice may awaken an idea, without itself being the product of an idea?, or that their vocalizations are not conventional?"[19]

I am aware that the observations that I made over the years cannot constitute sufficient evidence of the cognitive importance of vocalization in birds. The subject remains largely unexplored, and thus one is prone to error owing to the sparseness and the difficulty of observation. I do not lay claim to being always right. However, I hope that the reader will realize that birds are endowed with a capacity for communication based on vocal signs much more extensively than any other zoological class. Furthermore, they are able to make use of them to their advantage in different circumstances. Above all I am convinced that such vocal signs are the *unforced and willed* expression of sentiments, and that they are often used conventionally. We have investigated the physiological basis of the phonetic elements from which the voices and songs of birds are formed. Thus we have seen that, with very few exceptions, all patterns of vocalization in known birds, from the least talkative to the most talented singer, are compounded of a small number of simple timbres. Every species has a characteristic style of song that might easily serve as a marker for distinguishing it from other species. This peculiar fact is well known to experienced hunters.

The richness and variety of vocal expression in birds, despite the limited number of basic vocal elements, can only be explained by assuming that the development of infinite patterns of song is the result of psychic powers. This theory is opposed by those who follow Flourens in thinking that vocal expression in animals is forced and unwilled. But to rule out any psychic influence on birdsong, one would have to admit that it is possessed by the birds instinctively in all its variations since birth though possibly in a latent state. However, we shall presently see that, with the exception of the so-called nest voices, which in effect appear to be instinctive, birdsong is learned from the parents or from other birds of different species. In regard to the question of the conventional use of vocal signs, we will see that birds make use of them in order to call one another, love one another, challenge and mock one another, or convene together in certain places. A few examples will illustrate my views. On hearing an acute whistle common chaffinches become silent in our hunting cabins, song thrushes freeze suddenly on branches, and warblers take off disorderly, looking for shelter in nearby woodlands. It is clear that the whistle must have been perceived by them as signal announcing the presence of a bird of prey. When fighting with one another, sparrows, larks and robins sing in strange ways. Starlings, swallows, linnets, serins, ravens, and goldfinches cry out loud when they invite one another to congregate. Great tits (*Parus major*) make funny noises around the ludicrous postures of drowsy owls. Everywhere on the planet one can hear the passionate songs of the males of all species of singing birds that compete for the possession of females, courting them with their amorous talents.

Notes

1. {Du Pont de Nemours 1813, p. 236.}
2. *Historia animalium* Lib. VIII. Chapter 1. {Aristotle actually says: "...almost all animals present traces of their moral dispositions, though these distinctions are most remarkable in man. For most of them, as we remarked, when speaking of their various parts, appear to exhibit gentleness or ferocity, mildness or cruelty, courage or cowardice, fear or boldness, violence or cunning; and many of them exhibit something like a rational consciousness, as we have remarked in speaking of their parts. For they differ from man, and man from the other animals, in a greater or less degree; for some of these traits are exhibited strongly in man, and others in other animals". (Aristotle 1862, p. 104).}
3. Plutarch *Beasts are rational.* {Plutarch 1957, pp. 492 ff.}
4. {"Mirari solemus saltandi peritosquod in omnem significationem rerum et adfectuum parata illorum est manuset uerborum uelocitatem gestus adsequitur: quod illis ars praestat, hisnatura. Nemo aegre molitur artus suos, nemo in usu sui haesitat. Hoc edita protinus faciunt; cum hac scientia prodeunt; instituta nascuntur". ("We are regularly amazed at skilled dancers because their hands are able to represent all kinds of subjects and emotions and because their gestures are as quick as the works. What technique provides for them, nature provides for animals. No one has trouble moving its limbs; no one hesitates in making use of their parts. And they do so just as soon as they are born. They arrive with this knowledge. They are born fully trained)". English translation in Seneca 2007, pp. 88–89. While Seneca uses 'scientia' to indicate that animals come into the world already equipped with the know-how to move their limbs and parts, indeed already 'educated', as the language of 'institutio' may well suggest, Paolucci's introduction of a Christian theological category (*infused* science) in Seneca's discourse about animal cognition is striking. According to the Catholic Encyclopedia, "The existence of an infused science in the human soul of Jesus Christ may perhaps be less certain, from a theological point of view, than His continual and original fruition of the vision of God; still, it is almost universally admitted that God infused into Christ's human intellect a knowledge similar in kind to that of the angels. This is knowledge which is not acquired gradually by experience, but is poured into the soul in one flood" (Maas 1910).}
5. Montaigne *Essais.* {Paolucci must have in mind the *Apology for Raymond Sebonde*, see Montaigne 1893, pp. 154 ff.} Montaigne claims that a proximity to the divine condition is inherent in the very obligation to follow nature, to which the non-human animal is subject. In contrast he scorns the temerity and the fortuitousness of the freedom that we attribute to the human condition. "We enrich other creatures with natural gifts, and yield those unto them, that so we may ennoble and honour our selves with gifts purchased" (ibid., p. 154).}
6. Menault 1872, p. 97. {Cf. Menault 1868, p. 108. Paolucci cites the third edition, which appeared in 1872.}

7. Reimar 1770, Volume 2. {Paolucci interprets Reimar's complex analysis of animal instincts by emphasizing his negation of an identity of human and animal natures. This identity would entail the attribution to non-human animals of human traits such as reason, a possibility that Reimars wishes to rule out. Reimar's exposition of the theme of animal industrious instinct deserves closer scrutiny. He reduces the investigation of the mystery of animal instinct to the determination of fundamental forces that are conceived by him as the first principles of instinct, along the lines of the concept of force as determinant of motion in Newtonian classical mechanics. To this classical mechanical picture of the animal constitution he adds the notion of the animal soul's inner self-reflection, a capacity for turning inwards in order to apprehend its own nature. Perhaps an analogy put forward by him will clarify matters a little further. He states that a machine that is built like a clock would, by following the motive forces and the organization of its parts, be capable of one motion only. Next, he ingeniously suggests a thought-experiment. Let us conceive of the principal spring, or whatever else plays the role of first source of motion, of the *primum movens*, as the soul of the clock, and let us consider the other parts such as the balance-wheel, the gear-wheels, the hands, an so on, as organs by which the soul executes its own motions. Then, let us suppose that the first source of motion acquires a perception of its inner constitution, and let us imagine that the mechanical parts are replaced with organic parts such as senses, muscles, tendons, and others. This imaginary entity will be a living animal machine that, by following the inner perception of its fundamental force, which is essentially determined, will execute with pleasure and arbitrarily, albeit *blindly*, everything that it would have otherwise executed without sensation and like an artificial machine. It will no longer be restricted to one type of motion only, but it will regulate its action spontaneously on the basis of its external and internal perception. Cf. Reimar 1770, II, pp. 183–184, which I have followed very closely. I assume that this sentient yet blind modus operandi of the animal machine is what Paolucci perceives as Reimar's emphasis on the irreducible diversity of human and animal natures. In sum, a constitutive *blindness* regarding future-oriented action planning turns out to be, according to Reimar, the radical, incommensurable *otherness* of the non-human animal nature. The removal of this blindness would liberate the machine from its determined condition and hence elevate it into the realm of the human, where the soul's fundamental forces of motion are not determined but free to plan their own operations.}
8. {Paolucci does not give a reference for the citation from Nemesius. The passage quoted here is from *De natura hominis*, towards the beginning, where Nemesius (in Latin translation) says: "Nam huius pulcherrimae fabricae [i.e., the cosmos] opifex et factor videtur sensim dispares invicem naturas aptasse, ut unum essent et inter se cognatione quadam tenerentur. Ex quo maxime unum esse, qui universitatem rerum procreaverit, ostenditur. Non enim solum iunxit singulatim individual, sed etiam omnia, quae ubique sunt, apte inter se composuit. [...] Deinceps aedificator mundi Deus a stirpibus ad animalia transiens, non protinus ad gradiendi sentiendique vim devenit, sed gradatim eo et concinne progressus est." (Nemesius, Bishop of Emesa, 1802, pp. 2–3). The last portion of Nemesius's text evokes an image of the force of life and sentience being animated by the Creator throughout the cosmos by gradation, almost a diffuse process of awakening, in different degrees,

from the slumber of unconsciousness, during which organic life is gently raised into diverse states of conscious existence.}

9. Quatrefages 1861, p. 19. {Quatrefages presents an articulate, historically rich and philosophically complex view of the unity of the human species as opposed to the dis-unity of races that was often associated with racism and the scientific defense of slavery by anthropologists and politicians, especially in the United States (ibid., pp. x–xii). Although Quatrefages believes that there is a strong continuity between human and non-human animal in regard both to cognition and emotions and to their highest manifestation, namely, language, he finds the essential difference between them in what he terms 'morality' and 'religiosity', which, according to him, would be two spheres of eminently human endeavor that dramatically separate the realm of human beings from the realm of the non-human animals.}
10. Vignoli 1877. {Cf. Canadelli 2013, on this anomalous positivist scientist and his original research on animal psychology and the relation between myth-making in non-human animals and humans. Canadelli's book, "permette di comprendere che dietro ai nomi generici, come appunto quello di 'positivismo', si nasconde un ricco universo di pensiero che, nel caso di Vignoli, spazia tra filosofia, storia delle religioni, linguistica, etnografia, antropologia, psicologia, paletnologia, zoologia e fisiologia. Una varietà di prospettive che gli valsero, per esempio, l'ammirazione di Aby Warburg, che non esita a riconoscere l'influsso su di lui esercitato dagli scritti di Vignoli sul mito." (ibid., p. 5, from the preface by Elio Franzini). The *Archive Tito Vignoli* contains a wealth of information that has been digitized and is freely available at: http://www.aspi.unimib.it/collections/object/detail/8831/.}
11. Condillac 1766, Chapter 5, Part 2. {"Or l'instinct n'est que cette habitude privée de réflexion" (ibid., p. 109). But Condillac claims that although instinct originates in reflection, which is reason, there is an infinite difference between reason in humans and instinct in the non-human animals. "Il [i.e., instinct] est d'aiileurs infiniment inférieur à notre raison" (ibid., p. 112). In non-human animals, says Condillac, instinct is directed towards practical knowledge. It does not concern theory, as the latter presupposes method, that is, a system of signs that are employed to denote ideas, to order them, and to collect the results of these processes. In other words, he thinks that animal instinct judges only about what is good for the animal, what is practical, whereas human instinct judges not only about practical knowledge but also about what is true and what is beautiful (ibid., pp. 112–113). Thus, what is eminently human, according to Condillac, is judgment in the spheres of esthetics and theory-making.}
12. {Here Paolucci again cites *Ornitologia Italian*a, at p. 50, by Tommaso Salvadori, published in Milan, in 1872. The citation, I believe, might be an error for the reasons already mentioned above.}
13. Brehm 1869, p. 54.
14. Brehm 1869, p. 54. {Amazingly, according to Brehm, the parrot had learned how to form sentences that he had not heard, it asked itself questions and answered them, and was fully in command of the grammatical rules of time and number. Cf. ibid.}
15. Brehm 1869, p. 54.
16. The Swiss naturalist Heinrich Rudolf Schinz (1777–1871).
17. Brehm 1869, pp. 561 ff.

18. {This style of observational comportment by Paolucci, despite its obvious cruelty, can be fruitfully contrasted with Pierre Flourens's metaphysics, which rests on an ontological separation of human and non-human animals, and which Paolucci immediately criticizes (see below). Flourens (known for his disturbing researches on animal brains) uncompromisingly upheld the view that animals have some intelligence in the sense of skills and learned behavior but that they cannot have *reflection*, that is, according to him, the knowledge of thought by thought itself, which would be the demarcating trait of the human ("La *réflexion*, bien définie, est la *connaissance de la pensée par la pensée*. Et ce pouvoir de la pensée sur la pensée nous donne tout un ordre de rapports nouveaux. Des que l'esprit se voit, il se juge; des qu'il peut agir sur soi, il est libre; des qu'il est libre, il devient moral" (Flourens 1851, p. 67). In his opinion, the animal *follows the body*, whereas the human spirit, enclosed in a body, which clothes it with matter, is free, and indeed it is totally free, in the sense that, when it so wishes, it can immolate the body itself (ibid., p. 67). Flourens thus concludes with a quotation from Jacques-Bénigne Bossuet, according to which "the great power of will over the body consists in this prodigious effect, that man is master of his body to the point that he can sacrifice it for a greater good that he proposes to achieve. Throwing oneself in the midst of blows and succumbing to stabs because of a blind impetuosity, as it happens with animals, does not indicate anything beyond the body, while, on the contrary, to be determined to die knowingly and in accord with reason, notwithstanding the bodily disposition, which totally opposes this resolution, indicates a principle that is superior to the body. Among animals, man is the only one that manifests this principle" (in ibid., p. 68; cf. Bossuet 1875, p. 270). Paolucci's observational report of his comportment with the larks suggests otherwise. The mother does not hesitate to sacrifice herself in the face of the brutality of the hunter's act of separating the little ones from the nest. Nothing in the phenomenology of the episode hints at the lack of an infinite resolution in the animal's mind to die knowingly and in accord with her reason and with her own willpower over the body. But there still remains the possibility of a more radical difference. It is wholly conceivable that the non-human animal *will-to-life*, as opposed to the human-all-too-human *will-to-power*, can immolate the body in a style of comportment that need not subdue Bossuet's allegedly *reluctant* body, but in fact carries out the action in perfect and free *harmony* of body, will and mind. Only Bossuet's and Flourens's Christian theological prejudice seems to stand in the way of such radical possibility as the affirmation of the non-human animal difference. The animal discloses a non-human, cosmic dimension of will-to-life.}
19. Flourens 1851, pp. 64 ff. {"Mais enfin, ces voix, ces cris, ces accents, ces gestes ne sont que l'expression forcée, et non voulue des affections des bêtes. Ce n'est là, si je puis ainsi dire, que le langage du corps. L'esprit a aussi son langage où tout est artificiel, créé, convenu, voulu. Quand j'attache un mot à une idée, c'est que je le veux. Je puis le changer pour un autre. Si je sais vingt langues, j'ai vingt mots pour la même idée. [...] Le cri de l'animal peut bien réveiller une idée, mais il n'est pas le produit d'une idée. Et toute la différence est là. Les animaux ne font pas un langage; leur cris ne sont pas des signes convenus, des mots créé: ils ont des *voix* naturelles; ils n'ont pas de *langue*." (ibid., pp. 64–65).}

Bibliography

Aristotle. 1862. *History of animals: In ten books.* Translated by Richard Cresswell. London: Henry G. Bohn.

Bossuet, Jacques Bénigne. 1875. *De la connaissance de Dieu et de soi-même.* Nouvelle Édition. Paris: Librairie classique d'Eugène Belin.

Brehm, Alfred. 1869. *La vita degli animali, descrizione generale del regno animale. Traduzione Italiana dei Professori G. Branca, S. Travella, e altri naturalisti. Riveduta da Michele Lessona e Tommaso Salvadori. Volume Terzo.* Turin: Unione Tipografico Editrice Torinese.

Canadelli, Elena. 2013. *"Più positive dei positivisti". Antropologia, psicologia, evoluzionismo in Tito Vignoli.* Pisa: Edizioni ETS.

Condillac, Étienne Bonnot de. 1766. *Traité des animaux, où, après fait des observations critiques sur le sentiment de Descartes, et sur celui de M. de Buffon, on entreprend d'expliquer leurs principales facultés.* Amsterdam: Chez Ch. Ant. Jombert.

Du Pont De Nemours, Pierre Samuel. 1813. *Quelques mémoires sur différens sujets: la pluspart d'histoire naturelle, ou de physique générale et particulière.* Seconde Édition. Paris: De l'imprimerie de A. Belin.

Flourens, Pierre. 1851. *De l'instinct et de l'intelligence des animaux.* Troisième Édition. Paris: Librairie de L. Hachette.

Maas, A. 1910. *Knowledge of Jesus Christ.* In *The Catholic Encyclopedia.* New York: Robert Appleton Company. Retrieved from New Advent: http://www.newadvent.org/cathen/08675a.htm.

Menault, Ernest. 1868. *L'intelligence des animaux.* Paris: Libraie de L. Hachette et C[ie].

Montaigne, Michel de. 1893. *The Essays of Montaigne done into English by John Florio.* Edited with an Introduction by George Saintsbury. The second book. London: David Nutt.

Nemesius, Bishop of Emesa. 1802. *De natura hominis, graece et latine. Post editionem Antverpiensem et Oxoniensem, adhibitis tribus codd. Augustanis, duobus Dresdensibus, totidemque Monachiensibus; nec non duabus vetustis versionibus latinis, Cononis et Vallae, denuo multo, quam antea, emendatius edidit et animadversiones adjecit.* Edited by Christian Friedrich von Matthäi. Halle an der Saale: Apud Joan. Jac. Gebauer.

Plutarch. 1957. *Moralia. Volume XII.* Translated by Harold Cherniss and William Helmbold. Cambridge, Massachusetts: Harvard University Press.

Quatrefages, A. de. 1861. *Unité de l'espèce humaine.* Paris: Librarie de L. Hachette.

Reimar, Hermann Samuel. 1770. *Observations physique et morales sur l'instinct des animaux, leur industrie et leur meurs.* Two volumes. Amsterdam: Chez Changuion, and at Paris chez Jombert.

Seneca. 2007. *Selected philosophical letters.* Translated with an Introduction and Commentary by Brad Inwood. New York: Oxford University Press.

Vignoli, Tito. 1877. *Della legge fondamentale dell'intelligenza nel regno animale. Saggio di psicologia comparata.* Milan: Fratelli Dumolard.

PART TWO, CHAPTER TWO

Diversity in Birdsong according to Age and Sex

In the previous chapters we have studied the acoustic aspects of birdsong in relation to our perception. We have investigated the anatomical basis of animal vocalization particularly in birds. Now we need to approach the difficult question of purpose. We need to understand the function of birdsong as a means of social cognition in the life of birds. In other words, we need to inquire into animal psychology. In a biological act there always exists a psychic intentionality, in the same manner in which there exists physiological activity in any form of organic life.[1] In a fine work on linguistics, G. Dwight Whitney says: "The essential difference, which separates man's means of communication in kind as well as degree from that of the other animals, is that, while the latter is instinctive, the former is, in all its parts, arbitrary and conventional. [...] [N]o animal (that we know of) has any expression which he learns, which is not the direct gift of nature to him".[2] If it were true, this statement would defeat the claim of psychic intentionality in birdsong, and consequently research in this direction would be misleading. It would follow that in general animal vocalization could not be considered a language, no matter how rudimentary. This might be true, but for another reason that Whitney does not clarify.

In this chapter, I will try to show that while Whitney's claim might be upheld partially and under very special circumstances, in the majority of cases it turns out to be untenable and absurd. In subsequent chapters, I will discuss a series of facts that will cast light on the erroneous argument that birdsong is a *gift of nature to the bird*. However, I was relieved that Whitney himself, after declaring that animal communication is instinctive and automatic, added that "[i]t is, indeed, a matter of high interest, and promising of valuable instruction, to investigate as closely as possible the means of communication of the lower animals, so as to determine its character and scope; but the point calling for special attention is, how far the natural tones and utterances and postures and movements are used secondarily and mediately, for the purpose of signifying something".[3] This seems to moderate his previous claim that there is no analogy whatsoever between animal vocalization and human language.

In bird vocalization there is the instinct advocated by Whitney, but only before birds begin to learn and to imitate, that is, until their brain is sufficiently developed as to become capable of performing elementary cognitive functions. The period in the life of birds, during which they vocalize purely instinctively, is the infantile age, in the course of which they are educated by their parents. All birds in the nest possess an instinctive voice, or I might say a 'hereditary' voice, to use a less controversial terminology. The patient researches of Charles Darwin and Ernst Haeckel, to mention only the most important ones, have demonstrated the phenomenon of psychic inheritance in animals with as conclusive arguments as those on the basis of which organic inheritance has been considered as a fact for centuries. So, I need not list the fundamental facts that are relevant here. I will assume as granted that a bird can inherit from its mother its nest song, as well as its character, habits and specific type. However, this inheritance of song has a very limited scope in birds. When the bird becomes an adult and is no longer a nest bird, then inheritance ceases to operate. If the capacity for inheritance were unlimited, then Whitney's claim would be true. But I believe that the majority of bird vocalizations are not instinctive but rather neomorphic, that is, learned. To observe learning it will be sufficient to remove birds from their nest early in their life, when they possess the first voice only and do not have a brain sufficiently developed to remember the song that they might have heard from the parents. They must be raised either preventing them from hearing the song of other birds or giving them a teacher. In the first case they will continue to express the originary voice for the rest of their life. In the second they will learn more or less perfectly the voice of the teacher, according to the greater or smaller affinity that the teacher's voice has with that of the parents, especially in regard to timbre.

I removed from their nest two nightingales (the prince of singers) when they were still naked and possessed only that harsh and raucous accent typical of baby birds, which is maintained in both males and females for the entire duration of their life. I raised them in a closed room, without any chance that they might hear a song even remotely similar to that of the nightingale. They grew up almost dumb (except for the above mentioned raucous accent). I repeated the experiment with skylarks, which have the ability to imitate the song of many passerines. The outcome was the same as in the case of the nightingales. The birds grew up nicely and intelligent; they still live with me. They can only vocalize the nest accent, which they make when they wake up, or before sleep, or when I feed them, or when they are scared by the presence of an unknown object, that is, in all circumstances in which their relatives, which grew up and were educated by birds in the wilderness, would make use of diverse voices with a variety of articulations.

It is easy to raise a singing bird in the company of a teacher. Indeed this practice allows the observer to determine how far the capacity for imitation of the bird can reach. A male rock-thrush (*Turdus saxatilis*) learned only a short phrase that I taught him and quite different than that of the species. A blackbird learned a few musical stanzas perfectly, the calling patterns of the Eurasian skylark (*Alauda arvensis*), of the wheatear (*Saxicola oenanthe*), of the solitary thrush (*Turdus cyaneus*), which he then modified to his own purposes, using and abusing it especially during the love season. However, he never expressed a proper voice of his own species. I stole a couple of baby male black-headed buntings (*Euspisa melanocephala*) from their nest in the countryside. Both lost the nest voice. One of them learned perfectly the sexual song of the male ortolan bunting that I gave him as a teacher. The other could imitate the call of the finch, with which he had lived for sometime, only poorly. We will see later on the reason why the first succeeded while the second did not when we discuss the capacity for imitation in birds. These observations demonstrate, I believe, that most patterns of vocalization in singing birds are learned rather than inherited.

When a baby bird is hatched and comes into the world, it immediately feels the need to be helped by the parents, regardless of whether it was born in a comfortable nest, or marooned on a beach, or moorland, or desert. As soon as it is hatched, it will need the maternal assistance, on whose care its life depends. This is an inviolable law of nature, and it explains the impulse to communicate its needs, its suffering, its fear in the struggle for existence to its mother. It explains the use and the purposes of the nest song as a means of communication. Who has never heard the beseeching notes of a baby swallow, or of a finch, or of a nightingale, when they are taken from the nest where they used to receive their tender food from the loving wet nurse? Who has not seen the desolation of the chick when, after going astray in pursuit of insects, it suddenly realizes that it has been left alone and screams and calls the mother?

All birds have the nest song, which they use for some time while their life depends on the maternal care. An important comment has to be made concerning the nest song. If the adults are screaming birds, it has a different timbre than that of the parents. If the adults are of the whistling type, the song of their babies is similar to the wailings of human babies, and very often it has the timbre that will become the basis, or an important element, of the adult song. The simple patterns of the baby song are fused together in the formation of the voices that the bird will be able to learn afterwards.

One has to distinguish the differences in the infantile song of birds according as they are capable of self-feeding as soon as they are hatched, or *insessorial*, that is, incapable of feeding themselves, and hence fed by the parents. The former are born covered in feathers that protect them from the weather, to which they are immediately exposed. They feed themselves as soon as they are born. This type of birds is represented by fowls. These do not sing. They are crying birds. While crying birds generally do not whistle, their babies might be considered as whistling birds in regard to the voice. The character of the voice of these babies is totally different than that of the adults. Let us consider a familiar example, the domestic red jungle fowl (*Gallus Bankiva*), so that everybody will be able to replicate the observations that I will report. The raucous and harsh voice of both the male and female of this species are well known. But the cheeping of their chicks is rather different. It is doubtless a whistling sound. In the first days of life the chick makes the broken and syncopated notes that we are all familiar with, and with which it calls the brooding hen and responds to her calls. When the wings are covered in feathers, it can whistle at will, producing a sound long and trilling. The young chickens, both male and female, make this sound when they become aware of a danger, or something makes them suspicious. At age one month, more or less, if it is a cock the chick starts making the characteristic sexual call, but with a more whistling than crying timbre. Thus, I think that in the chick and generally in the nest birds that feed themselves, age triggers a transformation of the vocal apparatus, so that the whistling voice turns into a tracheaphonous one. However, this is only a hypothesis that I have made on the basis of analogy. More accurate observations will be needed to confirm it. What I said about chickens is applicable to geese, ducks, partridges, and all birds that feed themselves as they are hatched.

As to insessorial birds they are typically born naked, though occasionally covered in a thin layer of wool-like plumage, in a nest where they are cared for by both parents, or by the mother. After they are born they ask for food by making a characteristic sound of the species, which is truly instinctive and inherited. In general all singing birds are insessorial. In their babies the song has already the timbre and the sonic color of the adult song. But I was only able to make observations on this subject in indigenous birds. In the great tit (*Parus major*) the infantile phrase has the harsh timbre, according to the classification I suggested, which is heard in the special song of the male and in the adult song of birds of the same kind, as the blue tit (*Parus cocruleus*). In the baby green finch (*Ligurinus chloris*) the note is short and low, with the supple timbre, which is identical with that of the adult call. In the Eurasian skylark (*Alauda arvensis*) the song of the babies in the nest has the same trilling timbre as that of the father. In the goldfinch (*Fringilla carduelis*) the nest voice is a short, high whistle, with the acute timbre.

I have mentioned four examples referring to the four different kinds of timbres or fundamental sounds that I have proposed in order to show that in the infantile voice of singing birds one can find all the phonetic materials that compound the adult song. I have replicated the observations of nest songs in the blackbird, the skylark, the mistle thrush (*Turdus viscivorus*), the blue rock thrush (*Monticola cyanea*), the common rock thrush, the red-backed shrike (*Lanius collurio*), the common sparrow (*Passer Italiae*), the *Passerina melanocephala*, the chaffinch, and the goldfinch. In all cases I have noted the similarity of the nest voice with the song of the adult that has reached sexual maturity. But this similarity, or identity, mostly holds true only in regard to the color of tone, since the pattern, height, and intensity of the infantile song have proper characteristics. In rare cases the infantile voice becomes part of the adult song in all its musical integrity.

In general, the first caretaker of the babies, in birds as well as in mammals, is the mother. Baby birds take the food from the mother. As soon as they can fly, they follow the mother, albeit clumsily and while crying. It is the mother that comes first to the rescue and fights to defend them. It is natural to think that the song initially learned by the babies will be similar to that of the mother. Indeed, although the father, especially among singing birds, takes great pains in teaching the babies his sexual song since their birth, it is not until the love passion urges them to make use of all possible means to possess the female, that males abandon the song common to both sexes, which they likely learned from the mother. The song common to both sexes in all birds is made of both accents and musical phrases whose purpose is the cognitive communication necessary for the maintenance, prosperity, and survival of the individual. It is a valid instrument in the struggle for existence. In other words, it is the aggregate of all voices expressed by a species, with the exception of the song which is characteristic of adult males, which is produced in special circumstances, namely, during the mating season, and whose efficacy is principally reflected in the sexual competition.

If we consider the voices of birds in a single species in regard to both their acoustic patterns and the age at which they are used by the birds, we can establish three categories. 1) The song of the nest. 2) The song of the adults common to both sexes. 3) The song proper to the adult males. The first and the second have a function in the struggle for existence. They are useful in the same manner as beaks and talons are for birds of pray, or as tarsi and oblong feet if birds live in wetlands, or as webbed feet and flat beaks if they feed on food in water, or as the green color if they live in evergreen woodlands, or as the grey color if they live in moorlands, and so on. The third song is that of adult males. Together with many other means it serves the purpose of the sexual choice. It can be compared with the almost infinite series of ornamental organs of many male birds when they display the so-called wedding plumage, as well as with the splendid color of certain parts of the body, including the legs and the beak, with the development of special feathers in the tail (in peacocks, for instance), or around the waist (as in birds-of-paradise), or on the head (as in the *Acridotheres*). To this special song both naturalists and common people gave the name of 'song' (*ramage*) to indicate the rhythmic pattern expressed by it, which it is always possible to describe with a musical movement.

Ever since antiquity birdsong drew the attention of people, and it became clear that it is often an expression of love. Its biological importance is such that it deserves close scrutiny, notwithstanding the plethora of anecdotes that can be gathered from fiction writers, science popularizes, and poets. I will return to this topic in the next chapter.

In regard to the sonic importance of birdsong it can be stated that it is the most harmonious vocal expression that can be heard in a species of birds. The facility is astonishing with which the phonation and singing organs of birds lend themselves very rapidly to modulating the intensity, pitch, and timbre. While preserving the rhythmic structure, or the timbre, or the musical theme, certain birds are capable of surprising variations, which are amplified in myriad nuances through delicate articulations and harmonies. Probably these patterns of articulation correspond to the variety of psychic expression and thus form recognizable individual phrases.

According to Bechstein,[4] the nightingale can express twenty-five different songs. This learned German ornithologist translates them very well in syllabic notation (at least to the extent that syllables can resemble birdsong). They are as follows.

1. Tioû, tioû, tioû, tioû.
2. Spe, tiou, squa.
3. Tiô, tiô, tiô, tio, tio, tio, tio, tix.
4. Coutio, coutio, coutio, coutio.
5. Squô, squô, squô, squô.
6. Tzu, tzu, tzu, tzu, tzu, tzu, tzu, tzu, tzu, tzi.
7. Corror, tion, squa pipiqui.
8. Zozozozozozozozozozozozo, zirrhading!
9. Tsissisi, tsississisisisisi.
10. Dzorre, dzorre, dzorre, dzorre, hi.

11. Tzatn, tzatn, tzatn, tzatn, tzatn, tzatn, tzatn, dzi.
12. Dlo, dlo, dlo, dlo dlo dlo dlo dlo dlo.
13. Quio tr rrrrrrrr itz.
14. Lu lu lu lu, ly ly ly ly, liê liê, liê liê.
15. Quio didl li lulylie.
16. Ha gurr, gurr quipio.
17. Coui, coui, coui, coui, qui qui qui qui, gui gui gui gui.
18. Goll, goll goll goll guia hadadoi.
19. Couigui, horr, ha diadia dill si!
20. Hezezezezezezezezezezezezezezezezeze couar ho dze hoi.
21. Quia quia quia quia quia quia quia quia, ti.
22. Ki ki ki, ïo, ïo, ïo, ioioioio ki.
23. Lu ly li le lai la leu lo, didl ïo quia.
24. Higaigaigaigaigaigaigai guiagaigaigai.
25. Couior dzio dzio pi.

The chaffinch (according to Lenz, quoted in Brehm[5]) sings the following nineteen phrases.

1. Acute toast: zizizi willillilltih dappldappldappl de weingiche.
2. Bad toast: zizizizi illillillillillil sjibsjibsjibsjiwihdre.
3. Pine oil: zizizizi = rrrrezwoif zwoif zwoif zwoifidre.
4. Noise happy new year: tititititit etototototozespeuziah.
5. Happy new year from the Ercinia Woods: ziziwillwillwillwillsespeuziah
6. Ordinary happy new year: ziziziwihewihewihezespeuziah
7. Knight: zizizizizizullullulljobjobjobjereitjah.
8. Ride: zizizizirrrihtjobjobjobjobjerotihe.
9. A harsh dressing-down: ziziteuteuteutezelllllljoteuzipah.
10. Ordinary dressing-down: zizizizizizulleletscheutscheutscheuzipiah.
11. Warning: zizizizeuzeuzeuwillillillllwoifziah.
12. Whistle: zizizidisdisdisdisjibjibjibjibjaziah.
13. First extinguisher: zizizitolllelelelzwoifzwoifzwoifzihe.
14. Second extinguisher: zizizizitoitoiwillwillzihe.
15. Third extinguisher: disdisdistritritriclapclapclapzihe.
16. Fourth extinguisher: zizizillillillilltototozihe.
17. Thuringia song: zizizirrihtjibjibjibjiweidieh.
18. Embellished song: zizizizullullullullullullteufzziah.
19. Double bar: zizizizizizizizizizirrreuzipiah—totototototolazissssstutziah.

This variety of songs is not so rare among our woodlands and moorlands birds despite received opinion. It has been noted in chaffinches, in nightingales, and in a few other species, but this has been due to the particular attention that the melodic songs of these singing birds have always drawn. However, setting aside the melody, one has to admit that in regard to diversity and pattern variation, the robin, the tit, the whitethroat and many others are equal to the nightingale. By the same token, the songs of the ortolan bunting, of the linnet, and of the goldfinch are, in my opinion, equal to the song of the chaffinch.

Regarding psychic expression in birdsong I note that in general it is a means in the struggle among males for the possession of females. I will try to demonstrate this fact in the next chapter.

Notes

1. On the comparative study of psychic expression in animals, see the recent work by Tito Vignoli (1877).
2. Whitney 1876, pp. 338–339. {English translation from the original edition, Whitney 1875, p. 282.}
3. Whitney 1876, p. 348. {English translation from the original edition, Whitney 1875, p. 290.}
4. Bechstein 1858. {Cf. Bechstein 1838, p. 268. He describes the nightingale's song with musical language. "Nature has accorded both sexes the capacity for song. But the male has a particular talent for music. He surpasses all other birds in this regard, and hence the nightingale deserves the title of king of singing birds. The force of its vocal organ is admirable. The muscles of its singing apparatus are more powerful than in other species. But it is not so much the force as the extension, flexibility, prodigious variety, and harmony of the voice that make it unique to all those who have a sensibility for beauty. Sometimes it can elaborate on a single stanza composed of only two or three melancholic notes for minutes. It starts in a soft tone of voice, raising it gradually and reaching a superb crescendo at a peak of intensity. It then dies out. Sometimes the song takes on a more striking, ruptured rhythm, progressing through many repeated couplets, in ascending chords. One can count up to twenty-four stanzas or different couplets in nightingale's song without taking into account the minutest and most delicate variations. As it happens with all musicians so among nightingales there are great virtuosos and many mediocre singers. The song is so articulate, almost like spoken words, that it can be written down as speech. […] If we could understand the meaning of the song patterns [i.e., the twenty-five phrases listed above], we would find in them the expression of the most intimate sensations of this delightful singer" (ibid., pp. 267–269).}
5. {Brehm 1869, p. 155. Brehm states that the knowledge of these songs must count as a quasi-science, although they can be interpreted only by *the priests of this science*, for otherwise they remain obscure or even ridiculous to the lay people. In certain mountainous regions, according to him, this science is perfectly cultivated. Specialists and bird-keepers from

these areas can easily distinguish nuances where the untrained observer would be unable to discern any differences (ibid.). In other words, birdsong constitutes a special science, Brehm seems to be saying. Yet it is a science which has its own priesthood, and only the priesthood has access to the sacred knowledge of birdsong.}

Bibliography

Bechstein, Johann Matthäus. 1838. *Manuel de l'amateur des oiseaux de volière*. Déuxiéme Édition. Brussels: Société Belge de Librarie.

Brehm, Alfred. 1869. *La vita degli animali, descrizione generale del regno animale. Traduzione Italiana dei Professori G. Branca, S. Travella, e altri naturalisti. Riveduta da Michele Lessona e Tommaso Salvadori. Volume Terzo.* Turin: Unione Tipografico Editrice Torinese.

Vignoli, Tito. 1877. *Della legge fondamentale dell'intelligenza nel regno animale. Saggio di psicologia comparata*. Milan: Fratelli Dumolard.

Whitney, William D. 1876. *La vita e lo sviluppo del linguaggio*. Translated with anootations by Francesco D'Ovidio. Milan: Fratelli Dumolard.

PART TWO, CHAPTER THREE

Biological Purposes of Birdsong

With the word 'birdsong' (*ramage*) I will refer to the voice of adult male birds, especially the singers, which consists of more or less harmonious tones that are related to one another in a melodic pattern, which characterizes the song's structure. It is commonly heard in singing birds, and it reveals the artfulness of these gentle creatures. However, many *tracheaphonous* and *vocalizers* also sing although in an imperfect and simpler fashion. If one considers birdsong from a musical viewpoint, one notices that it always has a rhythm that manifests its underlying temporal structure. In regard to the timbre of the notes, I was unable to observe any regularity or norm that might determine the succession of harsh and supple tones or, say, that of acute and trilling ones. It is probable that this phenomenon depends on primitive processes still unknown to us and which may have regulated the early development of birdsong in different species. Sometimes the timbre of the supple tone predominates. It is the sweetest and most resembling to certain wind instruments. The song is harmonious and soft, such as in nightingales, blackcaps, robins, in many thrushes, green chiffchaffs, larks, orioles, and buntings, especially when the articulations of the different patterns that form the song accidentally follow the laws of harmony. This might happen, according to Horsfield, in the chestnut-capped babbler (*Timalia pileata*), whose song contains the five consecutive notes of the musical scale from E to B.[1] Sometimes the harsh timbre predominates, for example, in the sparrow, the starling, and the serin. Occasionally the song is a pleasant ensemble of different timbres, such as in the chaffinch, the goldfinch, and the linnet.

As regards the song of our singing birds it can be noted that it is generally sweet and melodic in warblers. It is more broken and less homogenous in ground finches. In both cases it consists of short notes, sometimes without pauses, sometimes with some delays (for example, in the chaffinch, the greenfinch, the siskin, and the bullfinch). This wealth of sounds distinguishes this type of song from others that are formed generally by very few notes (in the cock, for example), or by precisely three notes, such as in the goshawk (*Astur palumbarius*), which makes a simple *ghek ghek* during the love season. In a few rare exotic singers the harmony and structure of the song corresponds with the laws of music. Here is how Schomburgk and Waterton talk about the famous south-American bell bird (*Camarhynchus carunculatus*).[2]

> I was struck by strange sounds in the nearby wood that I had never heard before. I had the impression that invisible hands playing at the same time touched many glass bells in harmony. After a few minutes of intermission the sound would resume. A longer interval of six or eight minutes would follow before the sound started again. No others music could make me forget the fascinating singer.

In May, the female of the common quail sings a trilling but subdued *trichch* or *bribrib* (according to Brehm) , which is quite different than the resonant *quequerek* of the male. The female of the northern European snowy owl (*Nictea nivea*) expresses, when brooding, a peculiar *rik rik*, while the male makes a raucous *kra kra* (Brehm). According to Brehm, Gurney recounts that in a species of African toucan, the Abyssinian ground hornbill (*Bucorvus abyssinicus*), the male invites the female with a low *bù*, and the female responds by shifting the note one octave higher.[3] Furthermore, it seems that in domestic birds, since they need not struggle for their existence and are not subjected to the mechanisms of natural selection, the song of the female is influenced by domestication, while the psychic expression of the male song has been, so to speak, perverted. As we have seen, there are very few instances of special songs among the females of the best known species. In contrast, among our domestic birds, we easily hear hens singing like cocks, female canaries that trill like their males, and doves that coo.

Although the genesis of birdsong is generally unknown, in singing birds we can discern the basic vocal elements in isolated patterns used by the animals under diverse psychic circumstances and for different purposes. For example, in the case of the common sparrow, the song of the happy male alighting on roofs in the sunny days of early April, is an ensemble of voices of happiness, call, and warning, in which there is also mixed up a sexual reminiscence of the infantile voice.

The lark, which seems to evoke love with such great happiness in spring, you will hear trilling, suspended in midair, in the same manner in which she used to do in winter when warning her companions, or in fall when calling them, or again, as she used to do in the nest when she would eagerly ask her mother for a beakful of food. All these musical accents are merged together with artful variety, and often she mixes in her song the melodies of other singers that she imitates perfectly.[4]

The same can be said of the goldfinch, the linnet, the siskin, the wagtail (*Motacilla boarula*), the common house martin (*Hirundo urbica*), and others, even though the songs of these birds have a few characteristic traits that are not used separately by them. Thus, I would argue that some birdsongs are like summaries of virtually all the patterns of vocalization used by a species. They express one communicative need when they are fused together, but a single, different one when their components are sung in isolation.

In my view, birdsong expresses love in the vast majority of cases, and occasionally joy. That birdsong is most often the language of love in birds is suggested by the following facts. 1) The age at which singers express it with the greatest intensity. 2) The natural circumstances that are suitable to excite it. 3) The competition in which males engage in expressing it. 4) The cessation of birdsong during certain periods of the life of couples. 5) The relationship between vocal virtuosity and the display of color in the plumage.

I will presently discuss these five facts in relation to the observations which support them. Preliminarily, though, I would like to hint at some information that might indicate that in birds there also occurs a phonetic correspondence between male and female voices. In reference to the African boubou (*Lanarius Ethiopicus*), Brehm tells that the female responds to the clear song of the male consistently. In regard to the European black woodpecker, he comments that the male utters a thunderous *crrrrr* to which the female answers with a ***klik klik***. But these anecdotal observations cannot exonerate us from delving into a discussion of the above mentioned fundamental facts.[5]

1. The Age at Which Song Is Developed in Birds

So far, I believe, I have shown that birdsong must be considered as learned behavior. This is confirmed by the ignorance of the paternal song in young birds removed from the nest. The bird learns the song from the parent male or from other males in cases of social nesting. The fact that this learned behavior will be practiced at a much later time than the lessons received demonstrates that in birds memory is extremely powerful. It is well known that birds in the nest do not begin to vocalize their song until the second spring molt. However, in the first period of its life, the bird does not remain idle. It practices and does the homework in order to memorize the phrases that it will have to perform perfectly. I keep in cages many finches that I captured still young towards the end of summer. They remain utterly silent until the first molt, which occurs in September. Before that time they only possess the infantile voice. After about a month, if they have molted without much suffering, and if they have been kept in a warm environment with plenty of food, they begin to express the first notes of the song although the sounds are broken, syncopated, possibly wrong, and always very soft. However, it is difficult to distinguish these tentative patterns from the adult vocalizations common to both sexes, which have already developed. The ability to express the song in nest birds improves with age. It is almost unbelievable that this ability to sing perfectly is the fruit of great perseverance and practice since youth. In this regard I observed young males of mistle thrush (*Turdus viscivorus*), crested lark (*Galerida cristata*), greenfinches (*Ligurina chloris*), chaffinch, and goldfinch, and I always noted the following regularities. The bird usually practices after the meal. He perches on a stick in the cage, he makes himself comfortable, he ruffles his feathers and then starts practicing. He is very focused on what he is doing. He sings all notes, modulations, and rhythmic patterns in a very soft voice. It is clear that at the beginning he sings the different timbres with great difficulty. But very rarely does one notices a falsetto, a wrong note, or any other mnemonic error, even if one is an expert in recognizing all the delicate nuances of the adult song. The young artist proceeds with caution and prudently. When he rarely makes mistakes, he starts all over. He repeats the same stanza ad infinitum. This is the method with which he perfects the art of singing until he reaches maturity, that is, until he becomes capable of sexual reproduction.

The relation between the development of sexual power and song in the adult males is very obvious. They deploy the characteristic song when all their energy is dedicated to reproduction, which does not occur casually, when a male meets a female of the same species. In fact it always requires a more or less faithful and vigorous couple, which will be responsible for the offspring. It must be based on a moral understanding between the two birds forming the couple. I am convinced that birdsong is expressed willingly, as a sign of love, by the male, in order to pledge such moral understanding. The practice of song in singing birds was noted by Bechstein and by other French ornithologists, who call it 'marmotter', although I think that it has not been adequately interpreted and that it may have been misunderstood as mere vocal exercise.[6]

2. Natural Circumstances That Are Apt to Trigger the Song

Spring is the wedding season for all organized beings. Natural agencies powerfully influence sexual activity in plants and animals. Birds are no exception. As males compete for the possession of females, they deploy all the means at their disposal. I believe that song is the most powerful one. Heat is the main cause the triggers birdsong. This correlation has been noticed in birds that are held in captivity. Other factors are light and a calm atmosphere. The days of spring during which it is easier to hear most songs in most species in our countryside are calm and tranquil and with clear skies. If the temperature drops, or when the sky is cloudy and windy, the majority of our woodland and moorland singers will become silent. It cannot be doubted that heat is the primary cause that triggers birdsong especially when, aside from birds in captivity, one also considers birds wintering in our regions. We will presently see examples of both cases. I have had for many years a graceful canary that was silent during the day in winter, but which gave fantastic canorous performances in the evening when the room was heated with a stove. I have a nightingale that I captured when he was already an adult. He sings also during cold days in February as soon as he feels the warmth of a brazier. Other anecdotal evidence has been reported to me by worthy people. During the sunny days of winter, robins line up on the branches of elms, maples, and oaks, along the roads, and one can hear them sing their melodies with infinite variations and subtle nuances, for hours on end. Blackcaps and tits love duetting with robins while exhibiting their supple and harmonious voices.

3. The Singing Competition among Males

Charles Darwin mentioned an observation by Daines Barington Selborne (in the *Philosophical Transactions of London*, 1813), which, so far as I know, has not been repeated by modern naturalists. It concerns the way in which males emulate one another in singing.[7] I have already hinted at this topic when I discussed autophonous animals, especially batrachians. Now I would like to resume the argument in relation to birds. This will be important for my conjecture that birdsong is the language of love in birds.

Personal observations (corroborated by knowledgeable hunters that I have interviewed) suggest that the males of our migratory birds exhibit the song with greatest strength and most often a few days after their arrival in our regions. This is highly important. For, it is during the first days, when they have settled in their summer places, that they compete for the females. If they sing especially during that period, it means that song must have a special value in the selection process of the female. The question is whether song is used instinctively and unawares, or willingly. If the latter is the case, the value of the song would be understood by the bird. Consequently, it would be enough to demonstrate that this is indeed the case in order to show that song is a form of language, that is, a means of communication. The following observations are relevant.

At the beginning of May, when the ortolan buntings (*Emberiza hortulana*) come to our hilly regions and are busy nesting, I often noticed groups of males close to one female. All of the males sang with passion. But song was not expressed by them randomly, or unintentionally. I quickly realized that they were paying attention to one another's song. Indeed, the pause between one bird's song and another's was regular and constant. It seemed as if they were calling and responding to one another. Their voices had individual nuances of tone and articulation. To all intents and purposes, I was witnessing a singing competition among all the males gathered together in that place. I observed the same in the males of the corn bunting (*Emberiza miliaria*), which often gather together in large numbers. They were all perched on top of bushes. They repeated their unpleasant and harsh vocalizations incessantly while a few females flew around. The males of the cirl bunting (*Emberiza cirlus*) also sing together in competition. My father, who used to make painstaking observations of birds, told me that chaffinches do the same. During the period of passage, the nightingales sing marvelous melodies in the early morning hours. I could observe many of them in the same bush, which is more common when the males have formed couples with the females. I could report further observations concerning the chaffinch, the sparrow, the goldfinch, and the tit. But I will only mention the cock, which has some interesting characteristics relevant to our topic. The males are commonly heard competing with one another with the well-known *kiki riki*, to which each individual lends a special connotation, regardless of whether they are free-roaming or shut up in henhouses. Again, according to Brehm (*The life of animals*), two or three males of the wood pigeon (*Palumbus torquatus*) typically convene together to sing on branches. They keep going until the female chooses one of them, which suddenly stops singing and makes a subdued *pu pu* in sign of triumph.[8] Darwin received reports according to which singing birds, when they compete with one another, would have died of the excessive effort that caused the rupture of a pulmonary vein.[9] In the final analysis, all these observations support the claim that song in the males if often used as in competition for the possession of females, as a means of challenging the rivals, and as an endearing voice with their female companions. These conclusions, I think, are probably correct.

That a principle of competition among birds really plays a role in singing is made clear by the artificial means that excite song. I observed repeatedly that the rough noise of water rushing through the stones of a stream may elicit the song of the nightingale, of the common wren (*Troglodites europaeus*), and of other birds. A nightingale kept by my father used to start singing as soon as it heard the rustling of tissue paper grasped with the hands. Canaries, goldfinches, siskins, and linnets that we often have in our houses sing with passion when they hear people speak, or a spinning wheel, or the notes of a piano, and so on. In all these cases a principle of emulation might instinctively trigger the song. However, it is only a hypothesis. It must be considered with great caution until more detailed observations allow us to explain better why certain sounds often excite birdsong.

From the above considerations the conclusion might be drawn that a singing competition really exists among males that are aware of its usefulness for the possession of females. I might add another fact observed by others that concerns intelligence in singing birds. In the *Descent of man*, Darwin says that birdsong is an art that is improved by practice.[10] To substantiate his statement the great English naturalist mentions the ability of many birds to learn the song of other species or a musical melody. I will take up again this question in a more appropriate context when I discuss the principle of imitation. On the basis of what Darwin reported, I will mention how hunters actually select their decoys for attracting birds in snares. They do not take young males but only mature ones that have distinctive characters such as white forehead and large wings. These have practiced their song more than others and therefore sing with greater intensity and better intonation. They can also alternate their stanzas with great ease.

4. The Cessation of Birdsong during Certain Periods of the Life of Couples

We noticed that the greatest energy is expended in singing during the selection of females. Afterwards males continue to sing although less frequently and with declining passion, while helping the females build the nest and showing great affection at the time of brooding. When chicks are born adult males lose interest in the passion of love, which is replaced with the happiness of offspring. For a few days they stop cheering the woods with their songs. The courting language of the lovers turns into tender paternal accents. I observed this tendency in the nightingale and the chaffinch but the same is likely to be the case in other species. The nightingale loses his marvelous persuasiveness, and only answers with a sweet *ki ki* to the affectionate *crrrr* of the female. To express tenderness the chaffinch uses a trilling voice, low and soft, recognized also by Bechstein, who translated it as *trif*.[11] However, in these two species the disappearance of the song does not last long. As soon as the fledgling chicks open their eyes and maternal care becomes sufficient for their sustenance, the sweet voice of the nightingale and the happy beat (*Schlag* in German) of the chaffinch can be heard again. But in this case the song is not used for the possession of females any longer. It is used for imparting singing lessons to the offspring, as we have already noted, and perhaps also as an expression of joy. Thus, once they have reached their sexual objective and have secured mating, the males stop singing. This fact, too, suggests that the song is a means of communication among males and females concerning sexual reproduction.

5. The Relationship between Vocal Virtuosity and the Display of Color in the Plumage

Darwin illustrates the concept of selection and sexual struggle in birds with examples that are purported to demonstrate that the beauty of song has replaced the attractiveness of the color of plumage.[12] If we accept that birdsong is useful for these purposes, we must also assume that song has a communicative power that is understood by the male and seduces the female. This is the basis for discussing the intellectual merits of the song and of the other voices expressed by birds. I will return to these issues further on.

I would like to add other facts to those reported by the illustrious British naturalist, who mentions, among the European birds that display bright colors in their mating plumages but express only short and unmusical voices, the bee-eater (*Merops apiaster*), the common kingfisher (*Alcedo ispida*), the European roller (*Coracias garrula*), the Eurasian hoopoe (*Upupa epops*), the great spotted woodpecker and the green woodpecker (*Picus major, viridis*). He does not refer to others species and does not give examples of excellent singers that have modest plumages. But examples can be culled from the stupendous work on animals by Brehm (the *Illustrirtes Thierleben*), who never accepts accounts by other naturalists and travelers without a thorough investigation. Almost all species of parrots have beautifully colored plumages. Green predominates but other colors form harmonious combinations. Hues are rarely shaded. They are crystal clear, pure and stand out in marked contrast to one another. But the few voices of these birds are short, raucous, and screaming. One might say that they do not have song. It would have been replaced in its function by the colorful plumages. Let us remember the following family and genus, *Psittaculidae*, and *Strigops*,[13] which are well known for their concerts. I may add the *Cacatua*, the *Ara*, and the parakeets. The eastern rosella (*Platycercus eximius*) and the budgerigar (*Melopsittacus undulatus*) might be an exception, though the latter has the ability to imitate the song of others rather than possess a musical song of its own.

The trumpeter finch (*Bucanetes githagineus*) is a wonderful African sparrow, with bright grey and vivid red colors, but only expresses insignificant and melancholic voices. The Cardinal red bird (*Cardinalis Virginianus*), all vivid red with a lustrous black head, does not have a song (according to the Prince of Wied). The Australian *Amadyna* are beautiful and vivacious. They express low voices, unmusical and ventriloquial. Our rosy starling (*Pastor roseus*), one of the most beautiful European birds, has a poor, unmusical, graceless voice, which might be translated as *critsch*, *critsch*, *cirr*, *cirr*. The golden oriole (*Oriolus galbula*) has a bright black and yellow plumage but its voice is monotonous and finishes with a false note. Naumann transcribed it as *ditleô*, *gidaditleô*. Among the *Emberizidae*, which have a very modest plumage but are excellent singers, there are two exceptions in regards to voice and color. The black-head bunting (*Passerina melanocephala*) and the yellowhammer (*Emberiza citrinella*) do not have a melodious song but have bright colors.

The members of the admirable family of the paradise birds, comprising among others the *Paradisaea*, the *Cicinnurus*, the *Lophorina*, the *Seleucides*, the *Epimachus*, the *Astrapia*, have great beauty and their plumages are most vivid. In them nature wanted to sport luxurious plumes and silk streamers in delicate, wavy arrangements, creating glaring refractions and a changing play of light, but their song is infelicitous and unmusical. Many elegant members of the *Muscicapidae* have short vocal accents with little musical value, if any.

The Bohemian waxwing (*Bombycilla garrula*), among the most beautiful European sparrows, expresses its affection with a simple hissing sound, which can be imitated by blowing into a key (according to Brehm). The male of the *Ruticilla phoenicurus*, which is colorful during the mating season, has an inferior voice to that of his cousins. The same goes for the rock-thrush (*Petrocinela saxatilis*). The superb lyrebird (Menura superba) has a song similar to an out-of-tune ventriloquial sound. Many hummingbirds, small wonders covered in bright gold and precious stones, do not have song, which is a negative zoological trait, at least not one that can be reasonably be called 'song'. Among the European gallinaceous the heather cock (*Tetrao urogallus*) and the blackgame (*Lyrurus tetrix*) are most elegant, but they are really bad singers in their sexual competitions. The same is true of the jungle-fowl (*Gallus bankiva*), of the silver pheasant (*Nycthemerus argentatus*), of the *Phasianus veneratus*, and of the golden pheasant (*Thaumalea picta*). It is necessary to underscore, though, that not all cases support our views, as there are species in which sexual competition is controlled by both colorfully seducing plumage and melodious song. Among them we find the hard-bill (*Coccoborus Ludovicianus*), the goldfinch (*Carduelis elegans*), the northern mockingbird (*Mimus polyglottos*), and the Poe of Australia (*Prosthemadera circinata*). But these examples (and perhaps others that I am not aware of) may be few when compared with those species in which the colorful plumage has replaced the musical song.

Let us now consider facts of an opposite kind, which in nature occur frequently, but which lead to the same conclusions regarding the purpose and efficacy of birdsong, that is, cases in which song makes up for the lack of colors. Without going into details we may notice that the ablest singers of Europe have subdued or brownish plumages. Olive and grey hues predominate in the *Turdidae*, the *Luscinia*, and the *Sylviidae*. In the *Alaudidae* pale tawny and grey are the dominant tones, while greenish and ashen tints are most common in the *Fringillidae*.

So far we have dealt with birdsong as an expression of love. But this does not mean that it is always necessarily related to the sexual drive. As I have already hinted at, it seems that it is used by males also in order to express a feeling of happiness, a fact that can be observed in relation to both circumstances and the lack of association with females. Moreover, pure light, the sun, a bright horizon, and the transparent blue of the sky can all have an influence on birds, proportionately, as they also do on higher animals, by exciting the senses and reawakening the harmonious feeling of wellbeing that is enjoyed in the presence of nature. Under such circumstances I believe that birds employ song in the same way as human beings use interjections of admiration, or laughter and singing, which are patterns of linguistic behavior that can convey joy and love with great efficacy. In certain cases it is difficult to distinguish whether song is triggered by a feeling of satisfaction in the male or by the sexual impulse towards the female. I think that night songs are in this ambiguous category. I will never forget a beautiful summer night when I was walking through a copse wood in the Apennines. The moon shone in a limpid sky. The calmness of the atmosphere inspired the most exquisite sense of peace. Any creature would have been happy there. At that moment the happiness, which might have occasioned the noblest poetry in the soul of an inspired poet, must have been felt by myriad gentle creatures that resided in those woods. A solemn symphony of songs, whistles, garrulous notes, and bouncing echoes came from the bushes. The admirable musicians were the males of the nightingale and of the common whitethroat (*Curruca cinerea*). But how could we ever know whether those melodies expressed the happiness of existence rather than the lament of a troubadour?

Occasionally I heard red-winged thrushes (*Turdus musicus*), blackbirds, hedge-sparrows (*Accentor modularis*), and solitary sparrows (*Petrocincla cyanea*) singing in the night. According to the Swedish, the pine forest bullfinch sings in the tranquil summer nights. The same observation is reported about the South American hawfinch.

So far I have discussed birdsong almost exclusively in relation to singing birds. However, birds also display a variety of special voices during the mating season, although in different degrees, and on which similar remarks could be made. There is no point in minutely analyzing them. Thus, I will bring this chapter to a closure by offering a detailed survey of all the love songs that, I thought, lend themselves to being more accurately translated in syllabic notation.[14]

Notes

1. {Thomas Horsfield (1773–1859). Cf. Brehm 1869, p. 849, where the information reported by Paolucci, who does not cite his source, is discussed. Note that, according to Brehm, the notes identified by Horsfield are C, D, E, F, G. Paolucci may have misinterpreted the Anglo-Saxon nomenclature of music notation that was retained in the Italian edition of Brehm's work, which was consulted by Paolucci.}
2. {Charles Waterton and possibly Moritz Schomburgk. Cf. Brehm 1869, p. 786, for the quoted passage. Again Paolucci does not cite his sources.}
3. {Possibly John Henry Gurney (1819–1890), amateur ornithologist. Cf. Brehm 1870, p. 245, not cited by Paolucci.}
4. {This dynamic image of the flying lark, luminously evoked by Paolucci, calls to mind the highly original interpretation of the presence of this bird in European literatures put forward by Gaston Bachelard in his extraordinary book, *L'air et les songes* (Bachelard 1943, and English translation, Bachelard 1988). As Bachelard's prose is notoriously difficult to translate because of its musical and poetic force and rhythmic structures, which were masterfully employed by the French philosopher in order to reinforce the emotional engagement of the reader with his train of thought, I will quote a pertinent passage in the original, which offers insights into the imagery of the lark. It is a hermeneutic perspective that brings to bear the dynamic imagination investigated by Bachelard on Paolucci's poetic-naturalistic phenomenology of the lark's song. "Déjà, à quelques mètres du sol, l'alouette *poudroie* dans la lumière du soleil: son image vibre comme ses trilles; on la voie se perdre dans la clarté. Pou formuler cette éclatante invisibilité ne pourrait-on pas accueillir dans la poétique les grandes synthèses du génie scientifique. On dirait alors: *Dans l'espace poétique, l'alouette est un corpuscule invisible qu'accompagne une onde de joie.* [...] Et le philosophe, tout à sa fonction d'imprudence, proposerait une théorie ondulatoire de l'alouette. Il ferait comprendre que c'est la partie *vibrante* de notre être qui peut connaitre l'alouette; on peut la décrire dynamiquement par un effort de l'imagination dynamique ; on ne peut pas la décrire formellement dans le règne de la perception des images visuelles. Et la *description dynamique* de l'alouette est celle d'un monde en éveil qui chante par un de ses points. Ma vous perdrez votre temps à surprendre ce monde dans son origine, alors qu'il vit déjà dans son expansion. Vous perdrez votre temps a l'analyser, alors qu'il est synthèse pure de l'être et d'un devenir—d'un vol et d'un chant. Le monde qu'anime l'alouette est le plus

indifférencié des univers. C'est le monde de la plaine d'octobre où le soleil levant est dissous tout entier dans la brume infinie. Un tel monde a une richesse en profondeur, en hauteur, en volume, sans ostentation. C'est pour un tel monde sans dessin que l'invisible alouette chante" (Bachelard 1943, pp. 101–102). The lark is almost invisible, rising up in dazzling clouds of light, and can only be heard singing. The lark cannot traverse consciousness by means of ordinary perception, as visual images are impeded by the ascension of the bird in a glaring light, and thus sound becomes dissociated from the idea of springing from an inanimate sonorous body, or originating in a singing bird. The vibrating part of being, which is a mode of universal consciousness, is open to knowing the lark by responding to the bird's song in a resonance phenomenon, a tuning in soul between vibrating structures at the same frequencies. The vivid recollection of the lark that is presented in diverse literary accounts such as Paolucci's is mediated by the dynamics of the material, or *elemental* imagination delineated by Bachelard in *L'air et les songes* (and in other works). The element *air* is the substance of the dynamic imagination of flight and song. The lark embodies the dualism of flight and song, which is fascinatingly conceptualized by Bachelard in terms of the dualism intrinsic in modern quantum mechanics. The revolutionary physics of particles and waves was inaugurated by the emergence of quantum mechanical ideas in the 1920s and early 1930s, a process of which Bachelard was an original interpreter. The splendid analogy of the lark in terms of an invisible corpuscle that accompanies a wave of joy, that is, as analogous to a material point that in early interpretations of quantum mechanics and especially in the French tradition of wave mechanics championed by Louis de Broglie was literally thought as *accompanied* by a wave (which could manifest itself under special circumstances), introduces a powerful tension in Bachelard's analysis of the structures of the dynamic imagination. While the flight and song of the lark are continuous temporal phenomena occurring in an elemental substance, namely, air, which classically was considered to be continuous (Bachelard himself refers frequently to the alchemical tradition of the four elements, earth, water, air, and fire), the dualism particle-wave disrupts this classical framework by introducing in it a discontinuity (the *quantum* makes a discontinuous claim within the consciousness of reality). A creative tension between opposite conceptual frameworks was already present in many musical innovations at the turn of the twentieth century, especially in Claude Debussy, who introduced discrete timbre structures in his music, and on which Bachelard seems to have modeled his work on the material imagination. It is extraordinary that already in Paolucci's imaginative recollection of the lark's love song we find an emphasis on a *melody of timbres*, which points to radical developments in the esthetic of Western music. I am indebted to the work of Marie-Pierre Lassus (2010) for her inspiring interpretation of Bachelard's *musical* psychology of the elemental imagination in relation to Debussy's conception of discrete temporal and timbre structures.}

5. {Cf. Brehm 1869, pp. 723 ff., and 1870 pp. 63 ff., not cited by Paolucci.}
6. Bechstein 1840.
7. {Paolucci does not give a reference for this information taken from Darwin. See Darwin 1871b, presumably consulted by Paolucci, on p. 339, or 1871a, II, pp. 53–54. Darwin mentions Daines Barrington and White of Selborne. These two names appears to have been erroneously conflated into one single name by Paolucci. Moreover, Paolucci's reference to

the date of the *Philosophical Transactions* is wrong, the correct year (as reported in Darwin's work) being 1773.}

8. {Brehm 1870, p. 258.}
9. Darwin 1871b, p. 337.
10. Darwin 1871b, p. 340.
11. Bechstein 1838, p. 150.
12. Darwin 1871b, p. 341.
13. {Paolucci mentions two genera which I have been unable to identify. He calls them *Chrysomis* and *Piornis*.}
14. {The survey is presented by Paolucci as a summary table comprising a list of sixty-eight species. For each species a syllabic translation of the song is given, accompanied by one or two comments on the qualities of the songs. Since most of the examples are not first hand observations by Paolucci but are culled from other sources, I decided not to include the table. It seems to add little to the overarching argument. The reader may easily consult it in Paolucci 1878, also available online.}

Bibliography

Bachelard, Gaston. 1943. *L'air et les songes. Essai sur l'imagination du mouvement.* Paris: Librairie José Corti.

———. 1988. *Air and dreams. An essay on the imagination of movement.* Translated from Edith R. Farrell and C. Frederick Farrell. Dallas, TX: The Dallas Institute Publications.

Bechstein, Johann Matthäus. 1838. *Manuel de l'amateur des oiseaux de volière.* Déuxiéme Édition. Brussels: Société Belge de Librarie.

———. 1840. *Naturgeschichte der Stubenvögel.* Halle: Verlag Heyneman.

Brehm, Alfred. 1869. *La vita degli animali, descrizione generale del regno animale. Traduzione Italiana dei Professori G. Branca, S. Travella, e altri naturalisti. Riveduta da Michele Lessona e Tommaso Salvadori. Volume Terzo.* Turin: Unione Tipografico Editrice Torinese.

———. 1870. *La vita degli animali, descrizione generale del regno animale. Traduzione Italiana dei Professori G. Branca, S. Travella, e altri naturalisti. Riveduta da Michele Lessona e Tommaso Salvadori. Volume Quarto.* Turin: Unione Tipografico Editrice Torinese.

Darwin, Charles. 1871a. *The descent of man, and selection in relation to sex.* 2 vols. London: John Murray.

———. 1871b. *L'origine dell'uomo e la scelta in rapporto col sesso.* Translated by Michele Lessona. Turin: Società L'Unione Tipografico Editrice.

Lassus, Marie-Pierre. 2010. *Gaston Bachelard musicien. Une philosophie des silences et des timbres.* Villeneuve d'Ascq: Presses Universitaires du Septentrion.

PART TWO, CHAPTER FOUR

Adult Birdsong in Both Sexes

The love season in birds, no matter how fertile different species may be, is rather brief when compared with the time devoted by individuals to procuring their sustenance. Now, leaving aside the physiologic and communicative importance of song, we must recall that it is always one and the same psychic expression, and that, as we have seen, it may serve two purposes at most, namely, the expression of joy and of love. Let us also take into account the use of song in the training of offspring. Still, there remains a large number of vocalizations produced by both males and females that we will group under the general heading of 'adult birdsong in both sexes'.

When one reflects on the laws the govern the existence and evolution of the ornithological species in connection with the geography and climate of different regions, with their ecologies and large-scale meteorological and geological phenomena, and when one also keeps in mind the predisposition to flying, one will realize that the social life of birds must be immensely more complex than one might have suspected. Furthermore, it is very rare for birds to have perfectly solitary lifestyles. Indeed, I would argue that it is because of the social complexity of their lifestyles that their organs of phonation are so diversified and can serve as sophisticated means of communication and cognitive intercourse.

According to all ornithologists periodic migrations are a well established fact for a large number of species, not only in the temperate zone but also in the torrid zone and in the equatorial regions (according to the latest Livingstone journal).[1] These migrations are carried out by small groups of birds that have convened together before starting the journey, for example, the *Fringillidae*, starlings and larks, or by larger numbers of birds that communicate with one another especially when, during their travels, they are battered by storms. The rapid cooling down of the weather, or vice versa the hotter climate of the regions in which they reside for longer periods, often cause many species to migrate temporarily from mountainous regions to the planes and beaches, or from exposed, cultivated areas to cooler woodlands. Many predators make the birds to be in a constant state of alert. Any member of the group can play the role of guard, for example, in tits, chaffinches, blackbirds, and larks, or else the sentry can be a designated individual such as in sparrows, starlings, ravens, and cranes. Powerful stimuli for vocalization, that is, the production of sounds that express feelings, are satisfaction, happiness, cheerfulness, fear, anger, and triumph.

On the basis of these considerations, it can be suggested that the *language of birds* (if I may be allowed to use this locution) that is common to both adult sexes must have great psychic importance. The additional facts that I will presently discuss will hopefully confirm this conclusion. However, I must make a disclaimer. I have been unable to compare and contrast my observations with those of others except in a few cases. In consequence I have met with difficulties owing to both the lack of information needed to interpret the data and the impossibility of replicating certain experiments. Thus, although my overall argument appears to be supported, I acknowledge that in matters of detail the interpretation of the data might be contentious. I welcome future studies that might confirm or disprove what I have claimed.

It will be useful to group the voices that I will discuss according to their psychic origin, that is, according to their meaning, as follows. 1) Calling voices. 2) Warning voices. 3) Voices of surprise and fear. 4) Voices of joy and tenderness. 5) Voices of anxiety and anger. 6) Voices of meeting and concert.

1. Calling Voices

These voices announce the reciprocal presence of birds, or invite one another to get closer. Typically they are short, simple, subdue, and they are used by both birds that live in small groups in bushes, dry lands, or woodlands, such as robins, nightingales, blackbirds, thrushes, tree pipits, and by birds that live in large and compact groups (except during the mating season), such as larks, ravens, and starlings. Calling voices are the most commonly heard throughout the year by those who observe the behavior of birds in the countryside, even though they are used more frequently during the migration period. On a March morning I went hunting on the Adriatic coast. The air all around was calm and warm. It was at the crack of dawn, and the voices of birds that flocked towards the coast became more and more intense as they were reflected on the tranquil surface of the sea. Thrushes, mistle thrushes (*Turdus viscivorus*) and black birds insistently called one another. It was a chaos of echoes that filled the atmosphere. At end of October and beginning of November, during the migration period, it is common to see large scores of birds caught in storms at night while flying over the Adriatic. In the depths of darkness, battered by wind and rain, they get lost easily. One hears thrushes, black birds, larks, robins, lapwings, plovers, and other birds make their calling voices with great intensity. They answer and get closer when one imitates their calls, their trills and *zip*, their pathetic whistles. Almost every year in Ancona one can observe the same events. In 1874, the number of migrating birds caught in storms above our beaches was so great that scores of them were collected in the streets, drawn towards the city by the lights, where the poor animals ended up losing their bearings. It is well known that hunters use calling voices in order to make our woodlands and dry lands birds fly closer to them and alight on nearby trees. With a *zip* and a *tic tac* one invites thrushes, with a *zirl* the black birds, with a *tic tic* the robins, with a *ist ist* the meadow pipits, with a *bsiii* the tree pipits, with a *vitié* the parotias, with a *pio pio* the larks, with a *finck chiò, finck finck* the chaffinches, with a *trochióchió* the greenfinches, with a *pizpeck* the goldfinches, with a whistling *viù* the bullfinches, and so on many others.

One should notice that a calling voice imitated by humans turns out to be less efficacious or indeed ineffective during the mating season. At that time the sexual drive subdues the social instinct. In fact it antagonizes it, and birds flee from one another rather than flock together. By way of balance special voices are then used by both males and females to draw one another's attention. Examples are the soft, tender chirping of quails and larks, the barely audible *chi* of nightingales, a special trill of chaffinches, and so on. But these voices should not be confused with the songs that we have discussed above. Finally, calling voices are one of the most effective strategies for forming and maintaining groups among many of our birds that live together. One can easily observe larks, starlings, and chaffinches on branches or browsing together on the ground. Every time a member of the same species flies above, most likely one that lost its flock, it is invited by a bird in the group, and it immediately descends to make company with them in a friendly manner. During the March and October migrations I observed a migrating flock double its number in this way. The same is true for those birds that winter in certain regions in large groups, which the tricks of hunters can easily disperse.

2. Warning Voices

The many enemies that birds must guard themselves from led them to develop a formidable defense strategy in the struggle for existence. It is the use of special voices with which they warn one another of the imminent danger. They are most frequently made by males as they are typically on lookout duty. They are short and soft in small singing birds, raucous and screaming in certain larger species. Domestic fowl make a special call when a predator appears and hens run for cover. A chaffinch that is frightened by something unknown or by the presence of a hawk makes an acute whistle, on hearing which the group that was chirping together freezes silent, and if it is caught on dry lands it disappears in the closest woodlands. But if it is caught on branches, every member of the group remains still hiding behind leaves. Sparrows quickly hide in hedges, on roofs, in sheds, under stones, or wherever they see a hiding place as soon as one of them made the warning call with a special trilling whistle, in a low, guttural tone, repeated two or three times. Swallows that roam above the ground looking for insects fly upwards like mad when one of them makes a special call similar to that of sparrows, with which, as the hunters say, the bird *gives it away*. Among the numerous observations that made Brehm's encyclopedic work on the life of animals justly famous, I wish to emphasize especially those that he had carried out regarding the calling voices of certain birds that want to warn one another. He accurately reports the dampened *sii sii* whistle of the chaffinch, the guttural *tellterelltelltell* of the sparrow (which we Italians would translate better with the syllables *griò griò griò*), the harsh *tek tek* of the Eurasian blackcap (*Curruca atricapilla*), and the *sicsak sicse* of the spur-winged plover (*Hoplopterus spinosus*). I had long known the syllabic expression of these voices when I found out with great pleasure that Brehm's opinions were similar to mine.[2]

But in all the warning voices that I could study there is an important characteristic that distinguishes it from all other sounds made by birds. These warning voices are common not only to a single species but to many of them and often to a genus even though the songs are quite different. All species and genera understand that particular conventional sound. In other words, warning calls are a sort of special language, possibly primeval and fundamental in the struggle for existence, precisely because it was understood by different species and genera. The following observations support my contention.

The so-called *muffled* voice of the common chaffinch is commonly understood among the passerines by the *Pyrrhulae* (bullfinches, serins, and canaries), the *Fringillidae* (chaffinches, linnets, siskins), the *Coccothraustes* (greenfinches, hawfinches), and the *Emberizidae* (ortolan buntings, corn buntings, buntings), among the songbirds by all dry lands birds, such as *Humicolae* (nightingale, robin, blue throat), *Monticolae* (redstart, stonechat, wheatear, mountain thrush), by thrushes (*Turdidae*, common thrush and black bird), by the chiffchaff (*Phylloscopidae*, garden warbler and icterine warbler), by the wagtails (*Motacillidae*, yellow wagtail), by the accentors (*Prunellidae*, passerines, dunnocks, alpine accentors), and by the tits (*Paridae*, tits, chickadees, penduline tits, bearded reedlings, firecrests, and titmice).

The guttural, trilling whistle that we mentioned in regard to sparrows serves the same purpose in the tree sparrow (*Passer montanus*), in the *Petronia rupestris*, in the white-winged snow finch (*Montifringilla nivalis*), in the *Centrophanes Lapponicus*, in the tits, and when it has a sharp tone, in the rustic sparrow, in the *Chelidon urbica*, in the swift, and possibly in other birds that I could not observe accurately.

I often heard the *tek tek* of blackcaps being used as warning call by the *Troglodytes parvulus*, by the common shrikes (*Laniidae*, shrikes and woodchat shrikes), and by all the species of *Sylviadae* (warblers, red-eyed warblers). So far I have been unable to identify other warning voices aside from the *sii sii* (similar to that of the chaffinch), the *griò griò* (similar to that of the sparrow), and the *tek tek* (similar to that of blackcaps). This does not mean that there cannot be others that are used by European and exotic birds in the numerous orders of climbing birds, *Levirostres*, web-footed birds, and so on. I hope that others will contribute to this branch of learning that I think is of great biological importance.

3. Voices of Surprise and Fear

When they are in the imminence of dangers, birds have the feeling of fear like other animals, especially those that are commonly chased by humans. They express this feeling with special sounds, often highly pitched and repeated, like interjections in the animal language. They make this voice when they are put to flight by a bird of prey, or when caught by storms during migrations, or when ruthless hunters chase them down stealthily in woods or in thick bushes and in the dry lands. The voices of fear are heard precisely when the birds take flight. When the black bird and the common thrush are closely chased in hedges and made to take flight from bushes produce a series of short sounds, higher and higher in pitch, that out hunters call 'sgaggiare'. Quails trill in a soft tone when they are chased by dogs in thick meadows. Larks vocalize in special ways when they take flight from a nearby hill. Goldfinches make a subdued *pik pik* when they are just about to flee from a human being or an unknown object.

According to Brehm, under similar circumstances, the Eurasian jay sounds like *keh keh kreh*, the *Urocissa sinensis* like *pink pink pink*, the sparrow hawk (*Nisus communis*) like *ki ki ki*, the goshawk (*Astur palumbarius*) like a low-tone *vie vie*, the common house martin like a *scir scir*, the nightingale like a raucous *kré*, the Western orphean warbler (*Curruca hortensis*) like *rar rar*, the willow warbler (*Hypolais hortensis*) like a harmonious *tek tek terit*, the Carolina wren (*Thryothorus ludovicianus*) like a *cie cie cie … tirr tirr*, the *Orites caudatus* like a sharp *ziriri*, the common tit like a *pink pink*, the blue tit (*Parus coeruleus*) like a *zisteretetet*, the wryneck (*Yunx torquilla*) like a *scek scek*, the great spotted cuckoo (*Coccystes glandarius*) like a *kerk kerk*, the partridge (Caccabis rubra) like a sweet and sonorous *reb reb*, *scerb*, the *Sypheotidis bengalensis* like a stridulous and repetitious *cich cich*, the *Aegialithes minor* like a *dii, dii, dii*, rapidly repeated, the common snipe like a whistling *psip* followed by a sharp *scecci*, the *Stagnicola chloropus* like a *chirg, cherr, tett, tett*, the *Anser cinereus* like a *kakakak, kakak, kakakakakak*, the mallard (*Anas boschas*) like a raucous *reb reb*, and the little tern (*Sterna minuta*) like a repeated *crek crek*. All these voices of surprise and fear do not have biological importance except when they serve as warning calls for other birds, and therefore their use in the animal ecologies must have been restricted by nature especially for birds that do not live in close proximity of human beings.

4. Voices of Joy and Tenderness

We noted that aside from the expression of feelings of love, birdsong communicates joy, wellbeing, and the happiness that birds surely feel, for they are the freest beings sharing in nature's benign smiles. But song is a prerogative of males while the feeling of joy is common to both males and females. Thus, birds of both sexes make use of other voices that express their satisfaction. We should consider also the voices of tenderness that have often been confused with those of happiness. Notes of tenderness are often heard when birds have found large quantities of food and take pleasure in eating, or early in the morning, when dawn is tranquil and warm and the sky is clear, or in the hot summer days when they enjoy the fresh breeze near rivers and streams, or when they are busy nesting, brooding, and raising the chicks. These voices are generally harmonious, subdued and often include elements of song if the birds are singers. One of the most distinctive voices of tenderness and joy is that of the chaffinch. It includes a simple trill, sweet and harmonious, and is produced by males when they are caring for the chicks. The bird hunters of our Apennines know it very well. They call it 'morella' and it is a sign that chicks have hatched. The bird also makes a soft sound barely perceptible, repeated frequently when it is about to have sexual intercourse. Many woodlands and *Conirostrum* birds sing in a special manner, emitting a peculiar series of broken voices, when they spend time in the afternoon in bushes and on branches. The passerines that take flight with pleasure in the sunny days of April express their satisfaction with a melodic whistle that is reminiscent of their infantile voice. The same voice is heard when they have fun browsing in fields that have just been sown. Ducks are happy picking worms in the shallow waters of ponds. Hens chirp when the housewife gives them grains. I might add many more examples based on observations that I made on both domestic and wild birds. However, the biological pattern reveals the same structure and value aside from circumstances. Hence, I will draw this section to a closure by listing a few syllabic expressions (taken from Brehm or notated by me) that give an approximate idea of these voices in different species.

The voice of satisfaction of the cockatoos is a sweet *ka ka du ka ka du*, that of tenderness of the red crossbill (*Loxia curvirostra*) a short and feeble *gip gip*, that of the greenfinch a sweet *zrni zrni*, that of the corn bunting is a *tic tic*, that of the ortolan bunting a *ki ki* followed occasionally by a brief and sweet *pjhù*, that of the oriole a *bilov*, that of the common swallow a prolonged *vit* occasionally turning into a *videvit*, that of the *Muscicapa grisola* a *vistet*, that of the nightingale a *tak tak*, that of the blue-throat (*Luscinia svecica*) a sweet *fid fid*, that of the Orphean warbler a *bi ve ve vi*, that of the the tree pipit (*Anthus trivialis*) a light *ki ki*, that of the *Agrodroma campestris* a *critlin ztrlui zür*, that of the wagtail a subdued *quiriri*, that of the tit a *civui civui*, that of the hoopoe a deaf *queg queg*, that of the short-toed tree creeper a *sit sri si si*, that of the green woodpecker (*Gecinus viridis*) a *gich gich chech chip*, that of the European roller (*Coracias garrulus*) a feeble and high-pitched *kerh kerh*, that of the grey partridge a deep *curruc*, that of the quail a weak *bi-bi-vi*, that of the snipe a *diep*, that of the Eurasian curlew (*Numenius arquata*) a *tvi tvi*, that of the *Ardetta minuta* a dark *pumm pumb*, produced during the sexual intercourse, that of the goose a *tattatattatat*, that of the common shelduck (*Tadorna tadorna*) a *tiuioiaiuiei* (as reported by Naumann), that of the small tern a *kekerrek kikerek*, and that of the common seagull a *kek surr*.

5. Voices of Anxiety and Anger

Many birds produce voices with which they express frustration or pain, often anger, or even a sentiment of revenge and susceptibility, especially when they are sexually dissatisfied, exhausted by the struggles in which they engage with incredible energy, or are suffering the rigors of cold and inclement weather, or are starving. These voices are dark and melancholic, but sometimes acute and shrieking. One has to realize that they are more often and easily produced by aggressive birds and by those more prone to sexual excitement. The male rock dove (*Columba livia*), a kind of domestic pigeon, laments the absence of the female with a guttural *hûu hûu hûu*, which is promptly replaced with the song as soon as the female comes closer to him. When the sparrows engage in sexual battles in spring, blinded by rage and furor, they attack and topple one another on the ground, screaming with a loud *tell slip dis scilc-tell slip dis scilc*, clawed by the adversary's nails, or caught by the beak and hit by wing blows. The song thrush, the mistle thrush, the robin, and other wood singers have a particular voice, common to all, that can be rendered as a high and loud *trtrtrtr*, typically heard when they are caught in nets or by the hand of the hunter, or their wings have been restrained, or they find themselves in the throes of violent struggles.

According to Brehm, the snow chaffinch makes a pitiful chirp under the stress of anxiety. The goldfinches that in autumn vie with one another on the branches of thistles for their tasty fruits, accompany their pecks with a raucous burst sounding like *rererere*. Audubon (quoted by Brehm) recounts that the male of the red-winged blackbird (*Agelaius phoeniceus*) "displays great courage and tenderness. He protects the brooding female and attacks with horrible voices, which one might think express fear and expletives, any intruder that gets too close to the nest. He bravely confronts humans that come too close even though with no intention to harm the brooding birds. In case of danger he stays on branches close to the nest and emits such pitiful sounds that only ruthless people would remain untouched by his notes".[3] According to Neumann, swallows express extreme anxiety with a trembling *zec zec*. The spotted flycatcher complains with a raucous *ci rech tech tech*. The common rock thrush does the same with a *ùit ùit*. In regard to a water-thrush that Hoymer kept in his house, he said that "his death was moving. I had taken to give him food when he died in my hand while singing a feeble song". The dunnock expresses anxiety with a clear *di di*, the common whitethroat with a rapidly repeated *vik vik vik*, while the angered wryneck makes a weak *ved ved*, and the European roller sounds like *re re re* or even *rek*. The Lichtenstein's sandgrouse (*Pterocles lichtensteinii*) warms up for fighting by threatening the enemy with a *drod dro dra dre*. The Pallas's sandgrouse (*Syrrhaptes paradoxus*) produces a belligerent *cricricrik*. Starlings complain with a loud *rip rip rip rip*. The peewit moans with a *creit creit*. The woodcock signifies pain with a stridulous *cri cri*, and the curlew with a penetrating *creh* or *crih*. The Brent goose (*Bernicla torquata*) whistles when it is angry. The sea swallow complains with a soft *crech* that is quickly repeated in flight. Seagulls make a *cherechek ghirr* when they are furious, while the *Procellaria* sound like *karv*. These examples should suffice to confirm that birds can make use of particular voices to express anger or anxiety, in the same way in which they make use of different ones in order to call, warn, or express fear, happiness, and tenderness.

6. Concerts

We noted that all the types of voice that we mentioned, including the song, are used individually for a single purpose, by adult birds or by chicks in the nest, and by both males and females. However, in social birds or in birds that at least have a habit of convening together to rest, or during the night, this individual use of the voice is not the only one possible. Concerts are common at any time of the day. Our *Conirostrum* birds furnish interesting examples. In autumn passerines choose thick bushes and woods, or the banks of rivers to spend the night. Soon after sunset they flock together (or even earlier if a cloudy sky darkens), coming in small batches or one by one, and forming groups of thousands of birds. We observe the same pattern of behavior in young birds of the same species in August or September during the hottest hours of day when they get together in cooler places. Towards the end of summer, goldfinches convene in large numbers near streams, together with chaffinches, serins, yellowhammers, and tree sparrows. Both the nighttime and daytime congregations of these birds are accompanied by a spectacular concert, a prodigious mass of sounds, compounded of the songs, voices and many different notes made by all the participants. The concert of the passerines is indeed surprising, and it resembles the noise of a large waterfall.

I hid behind a thick bush to observe closely this extraordinary behavior. The noise bursts out suddenly when the orchestra has grown large enough. Every single bird on branches jumps hither and thither, fights with neighbors, and produces all sorts of sounds, i.e., the song, the infantile voice, the adult voices, the warning voices, trilling sounds, often receiving the pecks by nervous companions. It is a diffused, loud experience, a frenzy of sound, an indescribable sonorous feast. If the observer makes a noise the whole orchestra becomes silent and pays attention, only to quickly start all over if everything remains quiet. The same happens with the concerts of serins, goldfinches, and greenfinches during the hottest days of summer. But their performance is not as tumultuous as that of the passerines, and hence the concert is more in tune and agreeable. These are the voices that I have referred to as 'concerts'. Setting aside the question of their purposes, I think they suggest that the birds intend to sing together.

While Brehm does not discuss the examples I have mentioned, he describes the concerts of other species. The owl parrot (*Strigops habroptila*) of New Zeeland, very rare, has the habit of forming large groups that make impressively loud noises. He also reports that a male of the African great white-headed weaver (*Textor dinemellii*) "started his song with a *ti ti, terr, terr, terr, zerr, zeh*, and another male responded with a *gai gai zeh*, and a third with a *guik, guik, guk, guk, gheh*. Others yelled as loud as possible a *gü gü gü gü*, gheh. It seemed like a beehive, some came, others left, while I had the impression that all the young males were having a meeting on the tree".[4] According to Lenz, starlings tend to convene together from a great distance, in the evening, coming from everywhere. Towards the end of August, starlings that are scattered during the day get together in large numbers close to rivers, ponds, and lakes when the reeds are high. They fly in compact formations, or more or less numerous flocks, sometimes looking like dense clouds. They land on meadows, or reeds, at sunset, while screaming and making all sorts of noises, and bickering with one another, until finally they rest. The *Crateropus leucopygius* that lives in the woods of Abyssinia is one of the most impressive noise-makers and has the habit of forming groups that yell in formidable choruses. Bates recounts that toucans sing in a chorus when they alight together on tall trees. While one of them is the highest on the tree, conducting the orchestra, the others make inharmonious and dissonant sounds. Ducks and geese also make concerts that sound more like conversations when they get together close to ponds. The examples suggest that this habit of making concerts is common to all genera of birds, with the possible exception of prey birds. I have been attracted to these concerts for a long time. However, on the basis of my observations and the reports by others, I have been unable to form an argument about their purposes. Most of the times the concerts might seem an expression of happiness, a feeling of wellbeing common to all participants. When the concert is made in the place and during the time of rest, one might think that the sounds are the expression of anger in the struggle for securing the most suitable space in which they want to sleep. In these circumstances sparrows and starlings are indefatigable in challenging one another. But then it seems unexplainable that the concert is repeated when they wake up. Finally, one might think that these concerts give the singers a genuine musical satisfaction.

To summarize, on the basis both of observations of the multiple use that adult birds make of their voices and of the different circumstances surrounding their production, we can say that these voices have meaning both for the bird that expresses them and for the bird that hears them. Accordingly, the meanings will correspond to the sentiments that birds possess in so far as they are higher animals endowed with a complex nervous system.

Notes

1. {Most likely, Paolucci refers (without referencing the book) to Livingstone 1874, where numerous ornithological observations are reported, especially in regard to the habits of birds and birdsong.}
2. Brehm 1869, p. 923.

3. Brehm 1869, p. 306.
4. {No reference is given by Paolucci.}

Bibliography

Brehm, Alfred. 1869. *La vita degli animali, descrizione generale del regno animale. Traduzione Italiana dei Professori G. Branca, S. Travella, e altri naturalisti. Riveduta da Michele Lessona e Tommaso Salvadori. Volume Terzo.* Turin: Unione Tipografico Editrice Torinese.

Livingstone, David. 1874. *The last journals of David Livingstone in Central Africa, from 1865 to his death: Continued by a narrative of his last moments and sufferings, obtained from his faithful servants, Chuma and Susi by Horace Waller.* Two volumes. London: John Murray.

PART TWO, CHAPTER FIVE

Imitation Voices

Humans are the pinnacle of achievement in the animal kingdom. They are the only mammals that have been able to reproduce intentionally the voice of their relatives with organs of phonation. They have opened an infinite hiatus between themselves and the apes thanks to culture and the development of intelligence. This strange finding is all the more surprising when we consider that the opposite occurs in birds, which are all uniquely endowed with an extraordinary capacity for imitating the voices of their kin. Their imitation is perfect in so far as it replicates the voice of other birds. Most of the times it is virtually impossible to distinguish the genuine voice of a species from the imitation by another species. Those who have studied the behavior of birds will agree that this is indeed the case. Many observations confirm the great talent for the imitation of sound in birds. It is well known, for example, that parrots have such capacity. Many species of *Psittacidae* have this power of imitation even in the wild where they have not been exposed to human contact. In the presence of humans they learn how to imitate speech, which we will not discuss here. We will limit ourselves to considering the imitation of sound that occurs spontaneously in birds, and which can be of significance in the general biology of these animals. But let us note in passing that also ravens, starlings, magpies, and jays imitate speech.

The best known among *imitative* birds is the calandra lark (*Melanocorypha calandra*), which is very much appreciated by collectors for its ability. The song of this species is short and insignificant. But the imitative power of the calandra is extraordinary. These birds imitate the songs of others species to absolute perfection. Many times I heard a calandra lark imitate the voice of the chaffinch, the stanza of the common swift, the notes of the sparrow, the trills of the black bird and many other voices with high fidelity. In a family letter, Brehm's father reports that a calandra in his possession could imitate, entirely or in part, the song of the following birds: swallow, thrush, goldfinch, chaffinch, sparrow, wood pecker, tit, green chaffinch, quail, crested lark, skylark, heron, and in addition a whole host of songs learned from other singing birds that he had not been able to identify. According to Brehm, as

> the calandra exceeds in body mass other larks so it surpasses them in musical talent, and this is also true with respect to other musically talented species. The natural melody of the calandra is a cicada-like sound not particularly attractive, but the extraordinary fact is that whatever the bird hears is memorized in its imagination and subsequently reproduced harmoniously. In the countryside, the calandra is like an eco for all birds, and if one wishes to hear a variety of birdsongs one needs only to hear the calandra. The raucous sounds of birds of prey, stridulating voices, arias of singing birds, the calandra can learn them all and reproduce them in midair, a spectacular recital of arpeggios, phrases, virtuosity, and bravura. Exposed to the accordion it proves to be a superb student, and no other species can rival with its rapidity and extension and progress. Swiftly the calandra becomes itself a natural accordion, a living sound, vigorous, sweet, and indefatigable.[1]

Perhaps the black bird is second only to the calandra in terms of talent for learning and reproducing the sounds to which it is exposed. This is showed by the black birds raised very young in the nest and from which it is quite easy to hear entire songs. I own one that has been with me for ten years, and which learned very well the song of the skylark, of the *Turdus cyaneus* (Italian *passero solitario*, or solitary sparrow), snippets of the song of the nightingale, that of the redstart, and a few bars of a polka and of a military march. In other words, everything I taught him, the bird could reproduce with precision, not to mention what he had learned by himself, which included the weird noise of a shutter that I hear opening every morning.

Among the ablest imitating singers that are most commonly heard, I must indicate two species of warblers (*Curruca orphaea* and *Curruca nisoria*) and the willow warbler (*Hypolais hortensis*). When they are inspired, they perch on the highest leaf-less branches and hedges, and one can hear them sing with passion while enjoying the warmth of the summer sun. It is almost as if they wanted to produce a musical compendium of the voices of the tit, the swallow, the chaffinch, and the nightingale. Many treatises report the imitative feats of exotic singers. The northern mockingbird (*Mimus polyglottos*) ranks highest, and is believed by Audubon to be the king of singers. On the authority of Gerhardt and Wilson, Brehm reports the following stories about this bird.

> Songs vary according to place. In the woods, the mockingbird imitates woodland birds, in proximity of towns it mixes with its own song the sounds that are typically heard in farms, i.e., the clucking of hens, the quaking of geese and ducks, the miawing of cats, the barking of dogs, the squealing of pigs, and additionally the screeching sounds of door hinges, the noise of saws and mills, and countless other noises that it replicates with surprising effectiveness. Occasionally, domestic animals are taken in by its voices and are seen disoriented. Dogs wake up suddenly thinking that they have heard the voice of the master. Hens worry that they are hearing the chicks cheep because of danger. Domestic fowl are scared while mistaking the mocking bird for a hawk. The cat in love is cheated into thinking that the she-cat is nearby.[2]

The superb lyrebird (*Menura superba*) is a master of contamination and mixes its voice with that of others. Aside from the sounds of other birds, Becker noted that it could reproduce the barking of dogs and the crying of babies that it had learned from another member of its species, which lived in a firm in the nearby county of Gippsland. Other examples of birds that combine their voices with those of other species include the campo troupial (*Icterus jamacaii*), the *Cassieus cristatus*, the *Gracula musica*, the oriole, the red-backed shrike (*Lanius collurio*), the bluethroat, the black redstart (*Ruticilla titys*), the stone-chat (*Pratincola rubicola*), the common redstart, the Carolina wren (*Thryothorus ludovicianus*), and the *Calamodus phragmitis*.

It has to be noted that the power of imitation of sounds in birds has a relation to the structure of the vocal apparatus. In other words, the simulated loquacity of birds goes as far as the nature of the sounds to be replicated allows the mechanism of the larynx to function properly. This explains why the canary easily learns the song of the accordion. This instrument has a very similar timbre to that of the natural voice of the bird. I recall the case of two black-headed buntings (*Emberiza melanocephala*), one of which attempted to imitate its cousin (the ortolan, *Emberiza hortulana*) with great success, while the other reproduced the song of the chaffinch very poorly, as the latter has a quite different timbre than the *Emberizae*.

Now, intelligence must play an important role in these learning processes. The tendency to imitate, I would argue, is the psychic principle that awakens in birds the will to repeat the voices that they hear. Furthermore, in certain cases such as that of the canary, it is possible that an ability might be involved to conceive of musical harmony and to form a taste for it. In other cases it is possible that imitation is directed towards some other biological purposes that we do not know.

The tendency to imitate is present only in the order of passerines (with the exception of parrots), according to the classification of Cuvier, to which singers belong. The families that are most distinguished are the *Alaudidae*, and the dry land and woodland singers. I could not find other examples. This might imply that if imitation has a role in natural selection, it might turn out to be modest.

Finally, the specific song of imitative birds is generally brief, unsatisfactory, and insignificant (at least insofar as I can say, that is, in regard to the species I have studied). At any rate, this peculiar capacity of birds bears witness to a high degree of intelligence and memory. Very often a few lessons are sufficient to teach a phrase to a parrot, or an aria to a black bird, or an entire song to a canary. Quickly the imitation becomes perfect, complete and precise. The passion with which these artists pursue their talent is such that I have heard them repeat in dreaming what they had learned.

Notes

1. Brehm 1869, p. 281.
2. {Paolucci does not cite the reference.}

Bibliography

Brehm, Alfred. 1869. *La vita degli animali, descrizione generale del regno animale. Traduzione Italiana dei Professori G. Branca, S. Travella, e altri naturalisti. Riveduta da Michele Lessona e Tommaso Salvadori. Volume Terzo.* Turin: Unione Tipografico Editrice Torinese.

PART TWO, CHAPTER SIX

Conclusion: Phonetic Corruption and Regeneration in Birdsong

Important research has been carried out over the last few years since glottology rose to the rank of science. Its focus has been the origin, the development, and the life and decadence of human languages. It has confirmed that human speech is a natural phenomenon, which is accordingly proportional to the degree of human intelligence. It was born. It grows to perfection. Eventually it decays in the same fashion as the forms of organized beings develop and decay. Thus, the evolutionary laws discovered by the followers of Darwin and Ernst Haeckel can be applied to the study of the genesis of languages and dialects. Now, although the voices of birds can only to a certain extent be considered as means of cognitive communication among them, and hence as a language, I believe that the fundamental laws of human language can be applied to birdsong. This can be deduced from the many facts that have been discussed in previous chapters, in which we dealt with the phonetic elements of birdsong and its use. But more can be said that will confirm the analogy with human language.

Max Müller and many others think that the principal causes that determine linguistic change, and which are continually exerting their action, are phonetic corruption and regeneration. Their effects on the elimination and novel formations in spoken languages might correspond to the elimination due to non-use and the novel formations due to novel use that occur continually in the evolution of zoological species and plants. Along the same line of reasoning a language might be considered as a living organism in analogy with the life and evolution of plants and animals. In regard to the voices of birds, it is possible to recognize in their processes both phonetic corruption and degeneration. Thus, I believe that they can mutate and decay in the same manner as human languages. Corruption can happen because of accidental events, or because of a change in biological circumstances. Regeneration depends on the efficacy of novel sounds that are being introduced. Both phenomena are regulated by the structure and limitations of the vocal apparatus and can induce slight mutations in the nature of the voices of the same species, or they can change the color and character of the song in an entire family. Before reviewing the facts that support what I have suggested, it must be emphasized that, although these phenomena can be regarded as quite general, they can only be studied insofar as they emerge clearly enough to be appreciated and investigated in sufficient detail. Hence, for example, the phonetic mutations that occur in a species that has only a few, short voices remain elusive. In consequence, I have been able to study only the song patterns of the ablest singers.

Phonetic corruption and regeneration start influencing the song of a species by inducing acoustic differences in the songs of individuals. By listening attentively to the love songs of the ortolan, the tit and the robin, I noticed that they change slightly among males of the same species. The song is not reproduced with absolute fidelity in its tempo, pitch, timbre and movement. If the changes are minimal they disappear with the death of the individual. But if they are more significant they can be transmitted to the offspring. These mechanisms explain how a species of singers that is indigenous in a certain region has a different song than that of the same species from another region. Thus, the original language of that species gives birth to two or possibly more distinct dialects. The crested lark and the common chaffinch are two examples.

Our hunters know that the robins that inhabit the Apennines have in their song a special motif that distinguishes them from others that reside in the hills or close to the Adriatic. In reference to the chaffinches of Thuringia, Bechstein mentions that the song of those from the Hartz woodlands are different from those that live in the Rouhl region, or in the Voigtland mountains, or in the Langfeld, or in Austria.[1] Homeyer reports that the crested larks he observed in Spain had a quite different song than that in Germany and other countries.[2] I myself noticed a significant difference in the song of the larks of our hills from those of the mountains of central Italy. The phenomenon of regeneration naturally follows that of corruption and is an important phase of development in the life of a language. In this way a language acquires a number of distinctive characters, which are sufficient to determine its physiognomy and to separate it from others that are close to it, and with which it had a common origin. Sister languages are thus perfectly distinct even though they share a large vocabulary and the same grammatical framework, and in this respect they can be regarded as having been emanated from the same mother language. Now, in my view, phonetic regeneration and corruption have exerted their influence on the songs of different ornithological families though perhaps to a lesser degree than in human languages. These phenomena have established the confines and similarities, the affinities and family resemblances of birdsongs. I do not claim that birdsong can be compared with human language in a strict sense. But I believe that, since I have demonstrated that the voices of birds are elementary means of cognitive communication among them, they must obey the same general laws that have been discovered for human languages. All the voices of a certain ornithological species, which constitute the song of that species, have a degree of resemblance with those of another species with which they can have a family relation.

In order to track family relations in human languages, linguists have considered especially etymological phenomena, rather than similarities of nouns, adjectives, and verbs that can be rather superficial. Thus, they have established that, for instance, Sanskrit is the common origin of most European languages. In the same way, it is the timbre of the voices and songs of birds that must guide us towards recognizing the family relations, not so much the pitch, or the length, or the musical movement of the notes.

I will now furnish a few examples of the family relations among bird languages by focusing on the particular timbre that is the physiognomy of a species. Let us consider, among the tracheaphonous, the *Corvidae* (the crow family). In all species that I know of, namely, the *Corvus corax*, *C. corone*, *C. monedula*, *C. pica*, and *C. glandarius*, I always noticed the harsh, raucous, dragging timbre, which could hardly be confused with the screaming and trilling accents of the diurnal birds of prey, or with the sibilant voices of the *Laridae* (Gulls), or with the guttural timbre of the web-footed. Among the singers the observations are much easier and more certain than in the tracheaphonous. There is a marked difference between the voice of the *Conirostri* (such as conebills) and that of the *Fissirostri* (cloven beaks) and that of the *Tenuirostri* (such as slender-billed curlews). In the latter, soft, prolonged and melodious accents predominate. In the *Fissirostri* the voices are short like trills. In the *Conirostri* a mix prevails, more or less musical, which is formed by harsh, supple and acute or even trilling voices, although the voices are never as prolonged as those of the *Tenuirostri*. If we analyze in more detail the differences than we have done so far in reference to families, we can distinguish among the voices of the *Conirostri* those of the passerines, of the *Fringillidae* (Finches), of the *Emberizae* (Buntings), of the *Paridae* (Tits), and of the *Alaudae* (Larks); among the *Fissirostri*, those of the *Hirundines* (Sparrows), of the *Cypseli* (Swifts); and among the *Tenuirostri*, those of the *Turdidae* (Thrushes), of the *Humicola* (Canasteros), *Sylviae* (Warblers), and *Motacillae* (Wagtails). In all voices of these families there is a striking resemblance. Occasionally, in the same genus, one can distinguish them only by special accents or nuances in the tone, as for example, in the *Alauda pratensis* (Meadow lark) and *Alauda calandrella* (Small calandra lark), in the *Parus ater* (Coal tit) and the *Parus caudatus* (Long-tailed tit), in the redstart and solitary swallow, in the *Sylvia atricapilla* (Blackcap) and other Whitethroats, in the *Turdus pilaris* (Fieldfare) and *Turdus iliacus* (Redwing), and among the European white wagtails.

The relationships between ornithological languages can be ascertained on the basis of the resemblances of the voices, which become more and more evident as we consider orders, families, genera, and species. Thus, we can see that the genealogy of birdsongs corresponds to the genealogy of species. Consequently, the study of the voice is an important factor in establishing natural methods of classification. In some cases, it might bring us exact knowledge of consanguinity of species even more certainly than by following anatomical criteria. Consider the example of the rock sparrow (*Petronia stulta*). Until recently it was considered by ornithologists as belonging to the group of the *Loxiidae*. They took as criteria the anatomical features and especially the shape of the beak. Now it has been reclassified as belonging to the passerines, and indeed it resembles to them in the character of the song, while in this regard it is quite different from the green finch, the bullfinch, and the hawfinch. In the framework of our current knowledge of organic evolution, we can conclude that the comparative study of birdsong might further contribute to the understanding of the genealogical tree of birds.

I will bring this essay to a closure by emphasizing that, on the basis of our analysis of the structure, acoustic features and cognitive significance of birdsong, we can state the following two conclusions. First, birdsong is a rudimentary form of language that can be compared, as Lussana did, with the language of interjections, which constitute the originary, preformed materials of human languages. Second, the biological purpose of birdsong (among other means that serve this purpose though often less effectively) is the struggle for existence in the case of the song of the nest or of that common to both adult sexes, otherwise the purpose is sexual competition among adult male singers.

Finally, I hope that others will be able to continue and deepen the study of birdsong, which is a field of research that can advance our understanding of evolutionary theories, to which the progress of modern science owes so much.

Notes

1. Bechstein 1838.
2. {Probably from Brehm, but Paolucci does not cite the source.}

Bibliography

Bechstein, Johann Matthäus. 1838. *Manuel de l'amateur des oiseaux de volière*. Déuxiéme Édition. Brussels: Société Belge de Librarie.

Index

A

academic freedom and education
and pulp literature, 68
futility of, 75
limited in science, xii, 122
alchemy, 7, 141–143, 178
Aristotle, 123
and the genius of animals, 148
his *Historia animalium*, 152, 156
Aristoxenus, 35
Ascoli, Graziadio Isaia, 86
Astley, Thomas, 97, 98

B

Bachelard, Gaston, 177–178, 179
elemental imagination, 178
L'air et les songes, 177
the image of the skylark, 177
Bacon, Francis, 68, 78
Becker, Ludwig, 101, 102, 111, 195
Bagarotti, Giuseppe, 87, 96, 98
Bechstein, Johann Matthäus, 131, 132–133, 162, 164, 165, 170, 174, 178, 179, 198, 201
Beneden (van), Pierre-Joseph, 96, 98
Bernabò Brea, Luigi, 71, 78
Biblioteca Francescana, Falconara Marittima (Ancona), Italy, xv
biology
and laughter, 60
and teleology, 63
animal and plant, 121
biological jet engine, as a metaphor, 26
empirical regularities in, 115
existence of biological life, 122
mechanisms and cultural possibilities, 51, 52, 55
metabolism and resonance, 42, 59
moral equilibrium of biological life on earth, 37–38, 39
of birdsong's functions, 101
of hearing, 55–58

psychic intentionality in, 157
the law of functional substitution in, 115–116
Bopp, Franz, 86
Bossuet, Jacques Bénigne, 74, 78, 156
and the face of the beast, 30
Catholic Encyclopedia on, 28
in relation to Plutarch's *Gryllus*, 31
quoted by Pierre Flourens, 155
Boulogne, Jacques, 74, 78
Brehm, Alfred, 108, 110, 111, 112, 123, 126, 127, 131, 132, 133, 140, 142, 154, 156, 164, 165, 168, 169, 172, 174, 175, 177, 178, 179, 185, 187, 188, 189, 190, 191, 194, 195, 196, 201
Brunetière, F., 74, 78
Buffon, Georges-Louis Leclerc, Comte de, 60, 62, 77, 78, 148, 156

C

Calvert, James, 97, 98
Canadelli, Elena, 154, 156
Canestrini, Giovanni, 91, 96, 97, 98, 112, 113, 118–119, 121, 123
Capriolo, Alfonso, xv
Carlet, M. Gaston, 107, 112, 113
Caruso, Enrico, 40, 75
Cherniss, Harold, 71, 80, 156
Christianity
and imagination, 7
and modern consciousness, 30
and modern science, 28
in Della Porta, 21
its theology, 152, 155
related to scholasticism, 32, 37
source of Dante's philosophy, 13–14
Cicero, 70, 78
on Greek language, 5–6
Cioran, Emil M., 77, 78
Clapperton, Hugh, 98
Clark, Xenos, xii, xiii, xiv
cognition
birdsong as a means of social, 157
birdsong as an expression of animal, 147
complex in the presence of hearing and sight, 88
Dante on its workings in the heaven, 15
Flourens on, in animals, 24
in birds it suggests that they should be classes together with mammals, 150
in physiognomy, 20
language as an instrument for, 85
of beauty, 122
of sound, 56
organs of phonation in relation to, 101
Quatrefages on, in non-human animals, 154
related to resonance, 42
related to the unspeakable, 22
related to the will, 103
Seneca on, in non-human animals, 152
the concept of gradation in relation to, 149
the difference between human and non-human animal, 149
Condillac, Étienne Bonnot de, 149, 154, 156
Cook, James, 93, 98
Corsini, Bartolomeo, 111, 113
Corti, Alfonso, 44, 49, 55, 76, 78, 121, 123, 179
organ of, 45–46, 49, 57, 59
Cronus, 73
Cuvier, George, 86, 123, 196

D

Dalla, Lucio, 40, 75
Dante, (Alighieri), 62, 73, 77, 80
on *visage*/ will, 13–17
Darwin, Charles, xiii, 87, 88, 89, 91, 96, 98, 101, 108, 111, 112, 113, 123, 126, 127, 131, 132, 133, 158, 171, 172, 173, 174, 178, 179, 197
Debussy, Claude, 178
Della Porta, Giovan Battista, 39, 74, 78, 79
his role in the scientific revolution, 18–20

on the face, 30
physiognomy, 20–22, 27
Delamarre, L., 74, 78
Denham, [Dixon], 98
Derrida, Jacques, 70, 78
Descartes, René, 7, 75, 148, 156
diversity
among human languages, 35
ecological, 18
ethnic, 8
of human and animal natures, 153
of human character, 18
Donati, Piccarda, 15–17
Doron, Edith, 93, 98
Droste-Hülshoff, Annette von, 102
Du Pont De Nemours, Pierre Samuel, 141, 142, 143, 152, 156

E

ecological
argument, 65
diversity, 18
life-world, 68
mode of dwelling, 68
scale, 39
systems, 60
Eliade, Mircea, 67, 75, 77, 79
Eleusinian mysteries, 72
Epicurus, 86
esthetics, 11, 36, 50, 54, 91, 178
Eric (the parrot)
sings along with the tenor Luciano Pavarotti, 39–43, 46
timbre of its voice, 46

F

Fabricius, Johann Christian, 111, 113
Fascism
and Dante, in Italian education, 73
Flourens, Pierre, 74, 79, 150–151, 156
and Christian theology, 155
on animal cognition, 24–26
Foucault, Michel, 74–75, 79
on Greek *parrhesia*, 30–31
Fourier analysis, 44, 45–50, 56–60, 76, 121
Fourier, Joseph, 121

G

Gabrieli, Giuseppe, 19, 79
Galilei, Galileo, 7, 18, 71, 79, 122, 123
Girardi, Michele, 118, 123, 124
Gleich, O., 76, 79
Grimm, Jacob, 86

H

Haar, Alfred, 47, 76, 79
the Haar spectrogram, 47–48, 50, 53
Haeckel, Ernst, 91, 158, 197
Helmbold, William, 32–33, 80, 156
Helmholtz, Hermann von, 44–45, 55, 76, 79, 80, 117, 121–122, 124
and contemporary hearing research, 46, 48–49, 51
and theory of mind, 59
On the sensations of tone, 45
Huber, Pierre, 88, 96, 98
humanism, 22, 23
Humboldt, Wilhelm von, 86
Huxley, Aldous, 99
on folded textiles, 93

I

imagination
and Christianity, 7
and ethnicity, 73
and memory, 6

and physiognomy, 17–23
and spectrograms, 50
figures of, 6
Greek character of the, 10–12
in birds, according to Brehm's description, 194
in birds, according to Giacomo Leopardi, 62
in Du Pont De Nemours, 143
inflections of, 8–9
investigated by Bachelard, 177–178
Judeo-Christian, 7
non-humanistic, 7
poetic, 148
processes of the, in the biosophical realm, 4, 10
resonant power of images in the, 38–39
workings of, 5
imitation
capacity for, in birds, 159
expression and, in birdsong, 135–136
modest role of, in natural selection, 196
perfection of, in birds, 136, 193
power of, in birds, 195
principle of, 173, 193–196
related to curiosity and attention, 87
related to parrots, 127
intelligence
advanced in birds, 117, 150
Condillac on, in non-human animals, 149
Darwin on learning in birds as an example of, 173
demonstrated in animals by speech and mimicry, 96–97
denied to non-human animals, 148
De Quatrefages on, in non-human animals, 149–150
discussion of in non-human animals since antiquity, 147–148
equivalent in non-human animals to perceiving and willing, 24–25
evolutionary theorists on, in non-human animals, 149–150
Menault's account of in non-human animals, 123
Flourens on, in non-human animals, 155
in humans vs. non-human animals, 193–194
learning and imitation as, 195–196
related to memory, 196
Vignoli on, in non-human animals, 149–150

J

Jablonca, Eva, 51, 76, 79

K

Köppl, Christine, 76, 79
Kühne, Wilhelm, 86

L

Lamarck, Jean-Baptiste, 149
Lamb, Marion J., 51, 76, 79
Landois, Hermann, 103, 111, 113
Langemann, U., 76, 79
language
an organic entity, 86–87
defined by Paolucci, 85
G. D. Whitney's theories on, 125–126, 131
Giuseppe Bagarotti's discussion of, 96
Greek does not have an equivalent term for Latin *vultus*, 5
in biological life, 55
laws of human, applied to birdsong, 197–198
Max Müller on, 128–129, 132
memory, emotions, and body, 43
music as, 51, 76, 164
of empirical science, 33

of love in birds, 169, 171, 173
of noise, 112
of pitch and timbre, 47
origins of, 95–96
poetic, 67
problems with modern renditions of Plutarch's, 32
rational, 67
relation of to the unspeakable, 18, 20, 22
scientific, 68
possible sources of Paolucci's theories about, 96–97
trapped in dualisms, 37, 38
warning calls as, in birds, 184
lark
Bachelard on, 178
evoking love, 169
imagery of the flying and singing, 177–178
imitates the voice of the chaffinch, 194
poem dedicated to the, by Annette von Droste-Hülshoff, 102–103
Lassus, Marie-Pierre, 178, 179
laughter, xiii, 26–27, 60, 60–69, 77, 91, 127, 176
learning
ancient, (as culture), 18
and physiognomy, 20
in non-human animals, according to Plutarch, 68
in parrots, 127
language in humans, 131
observation of, in birdsong acquisition, 158, 194, 195
the rules of music, 29
to observe light, 9
Legée, Georgette, 74, 79
Lenz, Harald, 131, 133, 135, 142, 143, 163, 190
Leopardi, Giacomo, 67, 76, 77, 79
on birdsong as nature's laughter, 60–62, 67
on Plotinus's theory of nature, 63–66
Levi, Primo, 69, 70, 79
on human suffering in the Auschwitz concentration camp system, 2–3
Levinas, Emmanuel, 69, 70, 72, 79
le visage, 4–7
on useless suffering, 1
the infinite, 12
Lifschitz, Avi, 95, 96, 99
Livingstone, David, 182, 190, 191
love
anecdotal love story about the tenor Caruso, 75
birdsong, an expression of, 162, 169, 170, 176, 178, 181, 186, 198
birdsong, language of, in birds, 171
cat in, 195
expressed by insects in singing, 104
for the survival, 101
for offspring in birds, 150
in the poem "The lark" by Annette von Droste-Hülshoff, 102–103
social, in birds, 151, 159, 161
in reptiles, 108–109
instinct for, 79
love song of a bird translated in words, 148
mystery of, 90
nightingale's love song into French verses, 147–148
replaced with the happiness of offspring, 173
song of the gray-winged trumpeter, 126
Lubbock, John, 87, 88, 92, 97, 99
Lussana, Filippo, 96, 97, 99, 200
Lyell, Charles, 86

M

Maas, A., 152, 156
Mandolini, Giancarlo, xv
Mangani, Giorgio, xiii, xiv, xv
Marinelli, Giorgio, xv
Marzot, Giulio, 77, 80
mechanism
for frequency analysis in the organ of Corti, 45

of anatomical bellows, explaining the intensity of birdsong, 120
of biological hearing as resonance, 59
of hearing in birds, recent research on, 49
of phonation in mammals, 117–118
of sound in reptiles, 108
of sound-production in language, 94
of the Fourier spectrogram, 49
of the larynx, 195
of the piano keyboard, 45
philosophy of, xii, 7–8, 11, 34, 66, 68, 148
Menault, Ernest, 123, 124, 148, 152, 156
metabolism
and modernity, 22
and resonance, 42
mind, 3, 8, 10, 11
and the Haar spectrograms, 50
and scientists, 34
double death of humans and, 12–13
Greek concept of, 21, 28, 30
Helmholtz on, 46
human, 59
inexhaustible discussions about, 37
like a mental model, 26
modern heliocentric subjectivity of, 11
non-human animal, 155
numbed by prejudice, 95
of a bird, 48
of the parrhesiastes, 31
operates like a spectrogram, 48
modernity, 7, 22, 66–67
Montaigne, Michel de, 123, 148, 152, 156
Morbiducci-Mangani, Cristina, xv
Müller, Max, 86, 128–129, 132, 133, 197
music
and chastity, 68
and life-worlds, 143
animals like music instruments, 29
Greek theorists on, 35–38
human music-making, 54–55
in birds, 44–51
musical training, 76
notation of, 135–136
of the black field cricket, 105–106
of frogs, 109
of the great bittern, 126–127
of the savages, 97
perception of, 122
presence of music in the non-human animal kingdom, xii, 28
traditional music categories, 53
regarded as a language, 51
Myers, Charles, W., 112, 113
myth, 8, 12, 73, 154
and eschatology, 8, 12, 23, 73

N

Nardi, Bruno, 73, 80
Naumann, Johann Andreas, 131, 132, 133, 175, 187
Nemesius, Bishop of Emesa, 149, 153, 156
Nettl, Bruno, 42, 75, 80
Newmyer, Stephen T., 75, 80
Newtonian classical mechanics, 153
nightingale
cessation of its song during periods of time, 173–174
intensity of song in the, 120
its song described by Bechstein, 164
sings in winter in captivity, 171
song in captivity elicited by noise, 173
spectrograms of birdsong patterns, in the, 49–55, 61
variety of songs, 162–163
Ninnis, Elizabeth, 111, 113

O

Odrini, Guido, xiii, xiv
Ohm, Georg, 44, 58, 76, 80, 121
Oudney, Walter, 98
Owen, Richard, 86

P

Palmieri, Paolo, 71, 80, 84, 93, 105, 110, 120, 122, 124, 136, 137, 138, 141
Pascoli, Giovanni, 73, 80
Pavarotti, Luciano, 39–43, 46
Peteghem, Julie Van, 73
phoneticism
  as a dogma, 51, 58
  as human music and language, 55
  phenomenon of, 62
Pianigiani, Ottorino, 70, 80
Piazzini, Stefano, xiv
pitch
  and memory, 59
  and the musical qualities of birdsong, 44
  and time, 55, 58
  indeterminacy of, 129
  individual differences in birdsong, 110, 117
  in melodic sequences, 50
  in spectrograms, 40–41
  limitless variation of, in birdsong, 125, 128
  not objective, 47
  not salient in birds, 49, 199
  smearing of, 52
  perfect, 29
  rapid modulation of, in singing birds, 162
  related to bird anatomy, 95
  related to expression in birdsong, 135
  related to fear in birds, 185
  resolution of, 52, 55
  sensation of, 122
  the attribute of, in Fourier analysis, 48
  traditional music categories such as, 53–54, 76
  variation of, in birdsong reproduction, 198
Plotinus, 77, 80
  on nature's cretive silence, 62–65
  on the sleepiness of nature, 66, 67
Plutarch, 31, 34, 38, 71, 72, 73, 75, 80, 148, 152, 156
  The face on the disk of the noon, 7–13
  Beasts are rational, 27–34
Pokorny, Alois, 131, 133, 142, 143
poetry, xi, 15, 67, 77, 111, 148, 176
Porro, Alessandro, 123, 124
prosōpon 5–8, 10, 71, 72

Q

Quatrefages, A. de, 149, 154, 156

R

raven, 141, 143, 150
Reimar, Hermann Samuel, 148, 153, 156
Renaissance, 7, 9, 22, 23, 33, 70
resonance
  and frequency analysis, 46
  as sound-producing anatomical apparatus in birds, 105, 117–118
  between humans and non-human animals, 39–44
  modes of communication based on, 37–39, 178
  sympathetic, 29, 45–46, 59
rhythm, 4, 135, 164, 167

S

Salvini, Anton Maria, 122, 124
Sappey, Ph. – C., 119–120, 123, 124
Savi, Paolo, 131, 133, 142, 143
Sawyer, R. Keith, 42–43, 76, 80
science
  a form of neocolonialism, xiii
  Alfred Brehm on, 164–165
  and Christianity, 28
  and evolutionary theories, 200

and Giacomo Leopardi, 66
and Western rationalism, 37
as analytic form of reasoning, 18
as materialistic system of beliefs, 58
as symbolic system, 66
devastating influence of, on modern philosophy, 7
European, 34, 93, 122
influencing our thinking habits, 51
infused, 148, 152
its mechanistic dogma, 34
language as a, 86
of glottology, 197
of physiognomy, 20
of optics, 132
philosophers of, 34
popular, 162
related to Darwinism in Italy, xiii
related to knowledge, violence, ethics, and useless suffering, 3
related to myth, 8
related to philosophy, 131
related to positivism, 51
related to truth, 66

Selcer, Daniel, 66, 77, 80
Seneca, 148, 152, 156
spectrogram, 36, 40–42, 46–51, 52–60, 61
Speke, John Hanning, 98, 99
Stoppani, Antonio, 96, 99
subjectivity
and the infinite, 12
and the myth of Cronus, in Plutarch, 73
argumentative structure of ego-consciousness as, 35
demarcation of, 36
discursive, 9
imperialism of, 7
modern, 7
the logos of, 18
solar, 10–11

suffering
and anxiety in birds, 187
Flourens's animal experiments causing, 24, 33
human, in Giacomo Leopardi, 63, 66
in animal communication, 159
in animal decerebration, 27
in molting, 170
Levinas's phenomenology of useless, 1–2, 4, 69
Plutarch on animal, 33
Primo Levi on useless, 2–4, 70

Süßmilch, Johann Peter, 95–96, 99
syllabic notation of birdsong, 135–136, 138, 139–140, 141, 142, 162, 177, 179, 184, 187

T

Tasso, Torquato, 62, 77
Tega, Walter, xiii, xiv
timbre
and the larynx, 95
as tone color, 117, 125, 128–129, 136
four fundamental instances of, in birdsong, 129–130
in birdsong learning, 158–161
in Debussy's music, 178
limited in the *orthoptera*, 104
meaningless concept of, in experiments with humans, 76
modulated by the trachea, 118, and easily, 162
not subject to regularities in birdsong patterns, 167
of the accordion, 195
related to expression and imitation in birdsong, 135
reproduced with little fidelity in birdsong across generations, 198–199
sensation of, 122
variety and complexity of its structures in birdsong, 44–54, 59

Tonfarbe, 117
Turner, Steven R., 76, 80

U

University of Pittsburgh, xvi

V

Vanzolini, Paulo E., 112, 113
Vignoli, Tito, 149, 154, 156, 164, 165
visage, 70, 72
- and the infinite, 12–13
- as the face of the moon, 10–12
- Cicero on, 6
- Dante on, 13–17
- *d'Autrui*, 4, 5–7
- defined in the original French by Levinas, 70
- emblem of frailty, 15
- etymological questions regarding, 5
- Fascist ego opposes the, 68–69
- in the cosmos, 12
- in the non-human animal face, 6
- Levinas's philosophy of, 4
- manifestation of, 15
- mediating entity, 12
- not infinite, 17
- of the human/animal, 22–23
- of the lion, 21
- of the world, 22
- perception of, 15
- referred to as *vultus*, in Latin, by Cicero, 70
- related to physiognomy, 19–20
- related to will, 13–14
- reveals justice, 22
- revelation of, 17
- the beast has the, 30
- the epiphany of, 6

W

Wallace, Alfred Russel, 91, 96, 99
Whitney, William D., 86, 125, 131, 133, 157–158, 164, 165
Williams, Thomas, 97, 99

Z

Zanoni, Elena, 96, 99
Zeus, 73

www.ingramcontent.com/pod-product-compliance
Ingram Content Group UK Ltd.
Pitfield, Milton Keynes, MK11 3LW, UK
UKHW021827100926
8556IPUK00003B/204

9 781433 141065